Pull Production for the Shopfloor

SHOPFLOOR SERIES

Pull Production
for the Shopfloor

CREATED BY

The Productivity Press
Development Team

PRODUCTIVITY
productivity press

Productivity Press • New York

Most Productivity Press books are available at quantity discounts when purchased in bulk. For more information contact our Customer Service Department (888-319-5852). Address all other inquiries to:

Productivity Press
444 Park Avenue South, 7th Floor
New York, NY 10016
United States of America
Telephone: 212-686-5900
Fax: 212-686-5411
E-mail: info@productivitypress.com

Cover concept and art direction by Stephen Scates
Cover illustration by Gary Ragaglia
Content development by Diane Asay, LeanWisdom
Page design and composition by William H. Brunson, Typography Services
Printed and bound by Malloy Lithographing, Inc. in the United States of America

Library of Congress Cataloging-in-Publication Data

Pull production for the shopfloor / created by the Productivity Press Development Team.
    p.   cm. — (Shopfloor series)
  Includes bibliographical references.
   ISBN-13: 978-1-56327-274-5
   ISBN 1-56327-274-1 (pbk.)
   1. Production Management.   2. Production scheduling.   I. Title: Pull
Production for the shopfloor.   II. Productivity Press. Development Team. III. Series
   TS155 .P798   2002
   658.5—dc21

                                    2002005043

09 08   5

Contents

Publisher's Message

Pull Production for the Shopfloor is the fourteenth book to be published in the Shingo Prize-winning Shopfloor Series by Productivity Press. We have presented books on 5S, cellular manufacturing, quick changeover, mistake-proofing, just-in-time, kaizen, kanban, and several on TPM (including autonomous maintenance, focused equipment improvement, and overall equipment effectiveness). When you implement pull production, the various lean methodologies you have already learned and implemented unite to become a powerful totality for achieving and sustaining lean manufacturing. Before implementing pull production, you will want to have achieved a solid level of proficiency in several lean methods. But once you are ready for pull, this book will give you the process for getting there, and will present it to you in an easy-to-read-and-assimilate format. Throughout the book, you will be asked to reflect on questions to help you apply pull implementation to your workplace. Numerous illustrations will reinforce the text.

The introductory section, "Getting Started," will suggest reading and learning strategies, explain the instructional format of the book, and give an overview of each chapter. Chapter 1 presents the context for choosing pull production. To help you understand and prepare for pull, Chapter 2 discusses production management, stock and overproduction, push versus pull, and the requirements for achieving flow production. Chapter 3 tells how to establish an improvement infrastructure and takes you through the pull implementation process, step by step. Chapter 4 explains how to manage pull production with the kanban system. Chapter 5 discusses one-piece flow and advises how to bring suppliers into your pull process. Finally, a summary of the phases for implementing pull and resources for further learning are presented in Chapter 6.

In today's market, manufacturing companies will not survive without the ability to manage wide-variety, small-lot production while keeping costs down. Pull production gives you that capability. This book will show you how to bring your various lean efforts together to achieve pull production.

Acknowledgements

The development of *Pull Production for the Shopfloor* has been a team effort and we wish to thank the following people. Judith Allen, Vice President of Product Development, spearheaded this project. Special thanks to Diane Asay of LeanWisdom for shaping and writing the content, using material from the Productivity archives. Art Director Stephen Scates created the cover design and concept, with cover illustration provided by Gary Ragaglia of Metro Design. Mary Junewick was the project manager and copy-editor. Lorraine Millard created the numerous illustrations. Guy Boster created the cartoons. Typesetting and layout was done by Bill Brunson of Typography Services. Toni Chiapelli was our proofreader. Michael Ryder managed the print process. Finally, thanks to Lydia Junewick and Bettina Katz of the marketing department for their promotional efforts.

We are very pleased to bring you this addition to our Shopfloor Series and wish you continued and increasing success on your journey to lean.

Sean Jones
Publisher

The Purpose of This Book

Key Point

Pull Production for the Shopfloor was written *to give you the information you need to participate in implementing pull production in your workplace.* You are a valued member of your company's team; your knowledge, support, and participation are essential to the success of any major effort in your organization.

You may be reading this book because your team leader or manager asked you to do so. Or you may be reading it because you think it will provide information that will help you in your work. By the time you finish Chapter 1, you will have a better idea of how the information in this book can help you and your company eliminate waste and serve your customers more effectively.

What This Book Is Based On

BACKGROUND INFO

This book is about an approach to implementing pull production in order to eliminate waste from production processes, in particular, the wastes of overproduction and inventory. The methods and goals discussed here are closely related to the lean manufacturing system developed at Toyota Motor Company. Since 1979, Productivity, Inc. has brought information about these approaches to the United States through publications, events, training, and consulting. Today, top companies around the world are applying lean manufacturing principles to sustain their competitive edge.

Pull Production for the Shopfloor draws on a wide variety of Productivity's resources. Its aim is to present the main concepts and steps of implementing pull production in a simple, illustrated format that is easy to read and understand.

Two Ways to Use This Book

There are at least two ways to use this book:

1. As the reading material for a learning group or study group process within your company.

2. For learning on your own.

Your company may decide to design its own learning group process based on *Pull Production for the Shopfloor*. Or, you may read this book for individual learning without formal group discussion. Either way, you will learn valuable concepts and techniques to apply to your daily work.

How to Get the Most Out of Your Reading

Becoming Familiar with This Book as a Whole

There are a few steps you can follow to make it easier to absorb the information in this book. Take as much time as you need to become familiar with the material. First, get a "big picture" view of the book by doing the following:

How-to Steps

1. Scan the "Table of Contents" to see how *Pull Production for the Shopfloor* is arranged.

2. Read the rest of this introductory section for an overview of the book's contents.

3. Flip through the book to get a feel for its style, flow, and design. Notice how the chapters are structured and glance at the pictures.

Becoming Familiar with Each Chapter

After you have a sense of the structure of *Pull Production for the Shopfloor*, prepare yourself to study one chapter at a time. For each chapter, we suggest you follow these steps to get the most out of your reading:

How-to Steps

1. Read the "Chapter Overview" on the first page to see what the chapter will cover.

2. Flip through the chapter, looking at the way it is laid out. Notice the bold headings and the key points flagged in the margins.

3. Now read the chapter. How long this takes depends on what you already know about the content and what you are trying to get out of your reading. Enhance your reading by doing the following:

- Use the margin assists to help you follow the flow of information.
- If the book is your own, use a highlighter to mark key information and answers to your questions about the material. If the book is not your own, take notes on a separate piece of paper.
- Answer the "Take Five" questions in the text. These will help you absorb the information by reflecting on how you might apply it to your own workplace.

4. Read the "Summary" at the end of the chapter to reinforce what you have learned. If you read something in the summary that you don't remember, find that section in the chapter and review it.

5. Finally, read the "Reflections" questions at the end of the chapter. Think about these questions and write down your answers.

How a Reading Strategy Works

When reading a book, many people think they should start with the first word and read straight through until the end. This is not usually the best way to learn from a book. The steps that were just presented for how to read this book are a *strategy* for making your reading easier, more fun, and more effective.

Key Point

Reading strategy is based on two simple points about the way people learn. The first point is this: *It's difficult for your brain to absorb new information if it does not have a structure to place it in.* As an analogy, imagine trying to build a house without first putting up a framework.

Like building a frame for a house, you can give your brain a framework for the new information in the book by getting an overview of the contents and then flipping through the materials. Within each chapter, you repeat this process on a smaller scale by reading the overview, key points, and headings before reading the text.

Key Point

The second point about learning is this: *It is a lot easier to learn if you take in the information one layer at a time, instead of trying to absorb it all at once.* It's like finishing the walls of a house: First you lay down a coat of primer. When it's dry, you apply a coat of paint, and later a final finish coat.

Using the Margin Assists

As you've noticed by now, this book uses small images called *margin assists* to help you follow the information in each chapter. There are six types of margin assists:

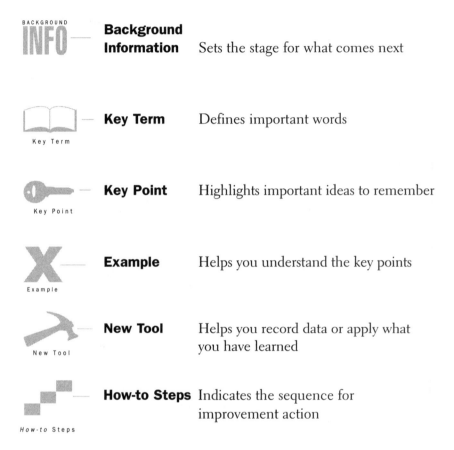

Background Information — Sets the stage for what comes next

Key Term — Defines important words

Key Point — Highlights important ideas to remember

Example — Helps you understand the key points

New Tool — Helps you record data or apply what you have learned

How-to Steps — Indicates the sequence for improvement action

An Overview of the Contents

Getting Started (pages xi–xvi)

This is the section you have been reading. It has already explained the purpose of *Pull Production for the Shopfloor* and how it was written. Then it shared tips for getting the most out of your reading. Now, it will present a brief description of each chapter.

Chapter 1: Choosing Pull Production (pages 1–13)

Chapter 1 provides a context for pull production and defines it. Next, it discusses some of the challenges of production and gives the prerequisites for implementing pull production. Finally, it lists the benefits of pull production to the company and individual employee.

Chapter 2: Understanding and Preparing for Pull Production (pages 15–34)

Chapter 2 describes the elements of production management and how they are addressed when shifting to pull production. The problem of stock and overproduction, push versus pull systems, flow production, and leveling are all discussed. The chapter ends with a brief description of where in the factory you should begin the pull implementation and gives the order in which to progress.

Chapter 3: Implementing Pull Production (pages 35–57)

Chapter 3 highlights the how-to of shifting to a pull system. It describes the improvement infrastucture needed and presents five key steps to implementing pull production. Level production, line balancing, and quality systems are also discussed.

Chapter 4: Managing Pull Production (pages 59–75)

Chapter 4 shows how to use the kanban system to manage pull. Six rules of pull production with kanban are given. The functions and types of kanban and how the kanban system works are explained. An analysis of pull production is presented, and one stage versus two stage systems, supermarkets, water beetles, and milk runs are described.

Chapter 5: Extending Pull Production (pages 77–93)

Chapter 5 discusses the final phase of pull, in which pull production supports one-piece flow. It reviews the conditions and steps for achieving production flow. How to extend pull production to your company's network of suppliers and subcontractors is also discussed.

Chapter 6. Reflections and Conclusions (pages 95–101)

Chapter 6 presents reflections on and conclusions to this book. It includes a summary of the steps for pull implementation. It also describes opportunities for further learning about techniques related to pull production.

Chapter 1

Choosing Pull Production

CHAPTER OVERVIEW

The Context for Pull Production

Key Point

Lean production or just-in-time manufacturing is a system of methods for eliminating waste in the production process. Every method in the system focuses on one or more typical sources of waste.

- *5S* focuses on waste resulting from disorder—scattered or missing tools with no designated places to keep them, poorly labeled materials and parts, poorly marked safety zones and storage areas, cluttered walkways, and so on—all those things that get in the way of the smooth flow of each operation.

- *TPM* focuses on the lost time and costs related to unplanned equipment downtime.

- *Quick changeover* focuses on decreasing the time needed to change equipment for different product specifications.

- *Mistake-proofing* methods eliminate lost time and costs related to unsafe procedures or procedures that result in defective products.

- *Cellular manufacturing* addresses transportation wastes that occur when equipment in the plant is organized by operation rather than by process; it enables mixed production and line balancing in relation to customer demand, as well as multi-task training for operators.

- *Kanban* helps to eliminate overproduction and work-in-process inventory. It is the key tool used within pull production.

- *Standardization* identifies standards for every operation and supports adherence to those standards until the next phase of improvement activity occurs.

- *Jidoka* (human automation) is the process of giving autonomy to operators to stop the line when defects occur and to eliminate sources of defects.

- There are some special conditions of pull production that address the wastes of WIP inventory, lead time, and idle time and waiting.
  - *Leveling*—the condition of small-lot, wide-variety consistent with customer orders.
  - *Line balancing*—shifting operators to lines where production is greatest.
  - *Multi-process operations*—the ability of operators to run multiple tasks or operations in their cells.

Key Points

Pull production brings all these methods together, and, together, they entirely revolutionize production management. Productivity Press has produced shopfloor books on many of these methods. *This book has been written specifically to explore the nature of production management within a pull system and the ideal sequence of establishing a pull system throughout a factory.* In Figure 1-1 the entire lean production framework is presented and the relationship of the various methods is shown. The highlighted methods will be discussed in this book as critical elements in establishing pull production.

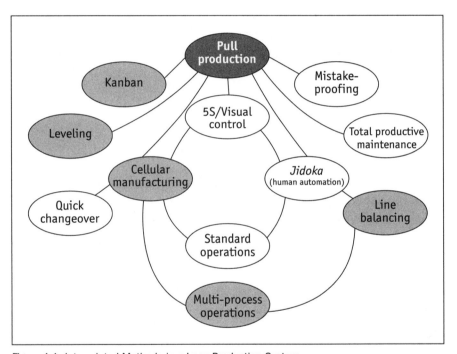

Figure 1-1. Interrelated Methods in a Lean Production System

We make a distinction in this book between cellular manufacturing and one-piece flow, suggesting that pull production depends on a cellular plant layout, but one-piece flow emerges most naturally once pull production has been implemented. Once implemented, the three methods work together as one. Implementing a pull system depends on awareness and at least partial implementation of all the other methods within the lean production framework, and a certain level of expertise among factory supervisors and operators of all these methods.

What Is Pull Production?

Two Aspects

Pull production has two aspects:

Key Term

1. *In manufacturing, pull production is the production of items only as demanded or consumed by the customer.*

2. *In material control, pull production is the withdrawal of inventory only as demanded by the using operation.* Material is not issued until a signal comes from the downstream user.

In a pull system the customer is the trigger for production and material withdrawal. Pull production is initiated by the *external customer* and triggered all the way back through the production process by the downstream or *internal customer* of each operation. It is a *market-in* approach to production.

Pull versus Push

Key Point

Pull production eliminates the waste that results from the more traditional push system of production, where material is moved from upstream to the next downstream operation as soon as it is available. In the push system, raw material availability is the permission to produce; and material procurement is based on forecasting customer demand. This is a *product-out* philosophy of production and results in over-production and/or delivery delays. To avoid the latter, inventory builds up in the warehouses and at each critical process junction. Bottlenecks occur where downstream processes cannot keep up with upstream production, and the pressure to produce results from upstream overproduction rather than actual market demand.

TAKE FIVE

Take five minutes to think about these questions and to write down your answers:

1. Who triggers production in the pull system?
2. How is this different from a push system of production?
3. What is a bottleneck? Do you have any in your factory? Name one and identify its cause.

The Challenges of Production

The Problem of Forecasting Customer Demand

Fluctuating customer demand for multiple product types and features creates a complex market environment within which to plan production output. These natural fluctuations make forecasting a complex process that must identify not only when demand fluctuations will occur, but also the trends in demand for specific product features—color, size, parts options, add-ons, and so on. Customer lead times must be considered as well as your own lead times; and unique customer specifications are probably impossible to provide, let alone predict. The pull system puts an end to this guesswork by making the actual customer order the trigger for production.

Figure 1-2. The Problem of Forecasting in Push Production

Small-Lot, Wide-Variety Production

Single-product factories are rare in today's world. Increasingly, factories must switch from one product model or feature to another. Dyes, drill bits and blades, and component sets need to be changed frequently at different points in the production process. Single-model, large-lot production no longer serves the demands being placed on manufacturers in today's highly competitive marketplace. Figure 1-3 shows the inventory and changeover costs of large-lot versus small-lot production. In the past, lengthy changeover times made large-lot production necessary. With pull production, the ability to reduce changeover times becomes a primary leverage point in lowering those costs and therefore increasing the feasibility of shifting to multiple-product, small-lot production based on market-in demands.

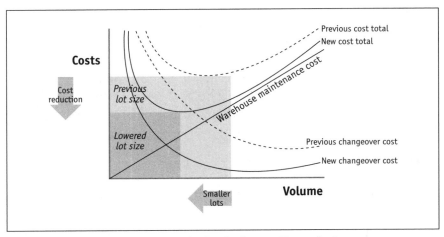

Figure 1-3. Changeover Improvements for Smaller Lots and Lower Costs

Cost Reduction

Key Point

Cost reduction is the primary aim of the Toyota Production System. The market-in, pull system approach to production defines cost differently in relation to profit and price than the traditional product-out, push system does.

Cost minus is the purpose of lean/JIT production.

In a product-out approach:

Costs + profits = sales price

In a market-in approach:

Sales price – profits = costs

Essentially these two equations express the same thing. There are three variables that remain in constant relationship to each other. However, in the first equation—the *cost-up* method—there is no incentive to lower costs. The sales price will cover them and, hopefully, the market will not object to the price you set. In the second version of the relationship—the *cost-minus* method, it becomes clear that you can only create price advantages by reducing your costs. If you want your profit margin to remain the same or go up, and you want your price to be competitive in the marketplace, you *must* reduce your costs.

Waste

Lean production targets waste as the critical factor in production costs. It is said that 80 percent of the cost of production can be found in process waste. Some of the essential ingredients to eliminating waste and cutting costs are shown in Figure 1-4. These are the focus of pull production.

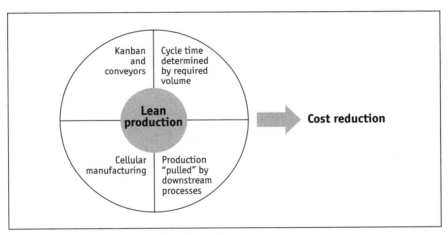

Figure 1-4. Lean Production and Cost Reduction

When Pull Might Not Be the Goal

If a company has only one product with no variations, or has seasonal and steep fluctuations in demand for that product, pull production will not offer as many benefits as it will for companies having a variety of products and product variations with customer demand constant throughout the year. This does not mean that process waste does not need to be identified; it just means that pull production may not be the ultimate goal of your waste elimination activities.

TAKE FIVE

Take five minutes to think about these questions and to write down your answers:

1. What is the primary cause of production costs?
2. How many product models and/or product variations do you produce in your factory?
3. How much space is devoted to warehousing raw material? Assembled products?
4. How much of the factory floor is used for work-in-process (WIP) inventories?

The Prerequisites of Pull Production

Before implementing the pull system you will want to be sure that you have achieved a certain level of proficiency in the first stage of lean, which includes the following methodologies.

An Awareness of Waste and Commitment to the Principles of Continuous Improvement

Operators, supervisors, plant management, and higher management of the company must all be aware of and dedicated to waste elimination and continuous improvement.

Team-Based Improvement Activities

Team-based activities to improve operations should be well underway and successful. There are many methods for guiding improvement activities on the shopfloor. Two popular ones are brainstorming and CEDAC (cause-and-effect diagram with the addition of cards). CEDAC is show in Figure 1-5.

Process Measures

Performance should be measured and rewarded on the basis of process improvement and the elimination of process waste.

5S—Visual Order, Displays, and Signals

All improvement activities should be recorded and the results displayed visually in the workplace. 5S should be in the final stages of implementation throughout the factory so that the workplace is

Figure 1-5. Team Members Using a CEDAC Board

well ordered and visual signals are in place. Operators must have the authority to stop the line when defects occur and andon lights/bells must signal the work stoppage so that instant help can come to resolve the problem.

Quick Changeover Activities

Bottlenecks due to slow equipment changeovers will continually emerge as you implement pull production and reduce the number of kanbans in the system. If methods for reducing changeover times are not being practiced, it will be impossible for the plant to respond to customer demand—the purpose of pull production.

A Multi-Task Trained Workforce

Union leaders must be involved in creating a training program for the workforce—one that will support the multi-tasking needed to balance lines in cells according to customer demand.

Cell Design

The factory layout should be based on cell design, where equipment is co-located according to process rather than function.

Figure 1-6 shows a sequence of five steps that can be adopted to establish a lean production system. Pull production, as discussed in this book, will concentrate on steps 3 and 4: cellular manufacturing, including multi-process operations and kanban; leveling; and some aspects of line balancing. More than with any other single lean method, the successful implementation of pull will require a synchronized, full-scale company commitment.

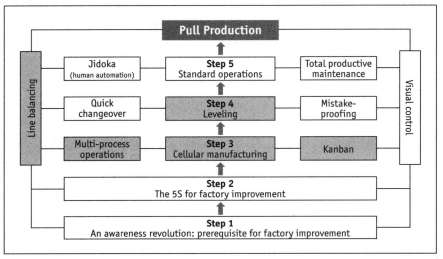

Figure 1-6. Steps To Establishing a Lean Production System

TAKE FIVE

Take five minutes to think about these questions and to write down your answers:

1. How many of the operators in your factory are involved in team-based, continuous improvement activities?

2. Have you implemented 5S? Andon signals? Process measures? Cell design?

3. Do you have learning programs in place for every operator to learn multiple tasks? Is the union involved in promoting this program?

The Benefits of Pull Production

Company Benefits

By adopting pull production, the company will achieve:

1. Powerful cost reductions.

2. Efficient utilization of the workforce.

3. Ease in identifying problems that need improvement.

Individual Benefits

Pull production methods can make individual jobs more satisfying because employees will:

1. Do work that is related to customer demand.

2. Gain increased skill levels.

3. Be empowered to improve the workflow.

In Conclusion

SUMMARY

Lean production or just-in-time manufacturing is a system of methods for eliminating waste in the production process. Every method in the system focuses on one or more typical sources of waste. The methods include 5S, TPM, quick changeover, mistake-proofing, cellular manufacturing, kanban, standardization, jidoka, and the special conditions of leveling, line balancing, and multi-process operations. *Pull production brings all these methods together, and, together, they entirely revolutionize production management. This book has been written specifically to explore the nature of production management within a pull system and the ideal sequence of establishing a pull system throughout a factory.*

Pull production has two aspects: *In manufacturing, pull production is the production of items only as demanded or consumed by the customer. In material control, pull production is the withdrawal of inventory only as demanded by the using operation.* Material is not issued until a signal comes from the downstream user. In a pull system the customer is the trigger for production and material withdrawal. Pull production is initiated by the *external customer* and triggered all the way back through the production process by the downstream or *internal customer* of each operation. It is a *market-in* approach to production.

Pull production eliminates the waste that results from the more traditional push system of production, where material is moved from upstream to the next downstream operation as soon as it is available. In the push system, raw material availability is the permission to product and material ordering is based on forecasting customer demand. This is a *product-out* philosophy of production and results in either over-production and/or delivery delays.

Cost reduction is the primary aim of the Toyota Production System. The market-in, pull system approach to production defines cost differently in relation to profit and price than the traditional product-out, push system does. If you want your profit margin to remain the same or go up, and you want your price to be competitive in the marketplace, you *must* reduce your costs. Lean production targets waste as the critical factor in production costs. It is said that 80 percent of the cost of production can be found in process waste.

Before implementing the pull system you will want to be sure that you have achieved a certain level of proficiency in the first stage of lean, which includes the following methodologies: an awareness of waste and commitment to the principles of continuous improvement; team-based improvement activities; process measures; 5S—Visual order, displays, and signals; quick changeover activities; a multi-task trained workforce; and cell design.

The benefits of pull production to your company include powerful cost reductions, efficient utilization of the workforce, and ease in identifying areas that need improvement. Benefits to the workforce include work that is related to customer demand, increased skill levels, and the empowerment to improve workflow.

REFLECTIONS

Now that you have completed this chapter, take five minutes to think about these questions and to write down your answers:

- What did you learn from reading this chapter that stands out as particularly useful or interesting?

- Do you have any questions about the topics presented in this chapter? If so, what are they?

- What additional information do you need to fully understand the ideas presented in this chapter?

Chapter 2

Understanding and Preparing for Pull Production

Approaches to Production Management

The increase in customer demand for wider variety in products and product features requires that manufacturers produce in small lots while still maintaining short delivery schedules and high quality. The old QCD formula is no longer sufficient; diversity in products must be added to quality, cost, and delivery excellence. Manufacturing companies will not survive without the ability to manage wide-variety, small-lot pull production. The ability to deliver greater diversity while keeping costs down requires lean production methods and, in particular, a pull production system.

What Is Production Management?

Let's examine the nature of production management to help us understand the purpose of pull production. Factories are living organisms with both information-based and equipment-based mechanisms that support the flow of goods to the customer. *Production management consists of these two important aspects*:

Key Term

- the information-based or "paper" system.

- the equipment-based or "physical" system.

The "paper" system includes the hard-to-see processes of planning, organizing, communicating, and managing the factory; it is often referred to as production control. The "physical" system includes the easier-to-see equipment, layout, conveyance methods, production methods, and other equipment-based factors through which raw material becomes transformed into product. The task of production management is to integrate these two aspects so that they function together effectively. Efforts to eliminate waste must address both aspects of factory life.

Computerization. The information-based aspect of production management is often addressed first—with computerization. But if the factory organization remains locked into the rules of large-lot mass production, these computer systems will be at odds with any changes you make in the equipment-based aspect of production management toward wide-variety, small-lot production.

Key Point

Computerized solutions for the "paper" side of production management have little or no impact on the "physical" aspect—the plant functions. *Once pull production is established, much of the information-based process will be eliminated or dramatically altered.*

Production Management

Production management means building and commanding:

an information based "paper" system (the organization, framework, procedures, information, management techniques, and other information-based organizing factors)

and

an equipment based "physical" system (plant equipment, layout, production methods, conveyance methods, and other equipment-based organizing factors)

to make effective use of people, materials, and machines to economically manufacture quality products just-in-time—just what is needed, just when it is needed, and just in the amount needed.

Figure 2-1. Production Management Defined

Consequently, computerizing information systems *before* making plant improvements will just add to the waste. You may be doing these operations more quickly, but how does that help you if you don't need to do them at all? Or if they are inconsistent with your market-in, pull system?

It's likely you already have computer systems in place to help you manage materials procurement and plan daily, weekly, and monthly production schedules. As you move forward in your implementation of pull production you will need to examine these systems carefully. Hopefully you will not confront too much resistance to the changes you will want to make on the "paper" side of your organization.

What Is the Significance of Lead Time?

Key Term

If we examine the issue of product lead time, the two aspects of production management come into focus. *Product lead-time begins with the production plan and ends with product shipment.* Figure 2-2 shows the various elements included in product lead time in a typical push system.

As you can see there is much more to the paper side than the physical side of a production plan. One does not move directly from sales planning to building products. Production capacity

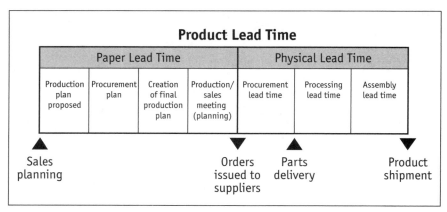

Figure 2-2. Product Lead Time In a Push System

must be considered and the plan must be tailored to meet the needs of both production and sales. Then it must be written as a formal production schedule, including delivery schedules for assembled components, parts, and materials needed to meet the sales forecast, and instructions for procurement and subcontractor orders to meet the production requirements. Throughout this aspect of the product lead time, no physical production has occurred to meet the plan. It is "paper lead time."

Only when production orders are issued do materials begin to flow. First procurement and sub-contractors respond to orders for materials and parts; and deliveries arrive from them within their prescribed delivery time. The factory gears up to produce the sub-assemblies; and finally parts are assembled into finished products and delivered to the customer.

It is obvious why large-lot production and warehousing emerged from the mass production model. Customers will not wait through the *entire* product lead-time to receive their product. Consequently, in a push system, there must be sufficient finished product and sub-assemblies stored to meet future possible demand within the customer's needed delivery time. In this system the lead-time is so complex and lengthy and dependent on so many external processes that the arrival of materials and parts becomes the trigger to manufacture products.

The cost of doing this has been ignored until recently; and the solutions implemented by only a few—who have become leaders in the global marketplace. The ability for actual customer demand to trigger production is already the competitive factor in today's economy. Shifting to a pull system must become the priority for any manufacturer who would survive and thrive in the emerging world markets, where the cost of process waste will be the difference between having satisfied customers or having no customers at all.

How Does Lean Production Impact Lead Time?

The factory-based improvements addressed by lean production methods are the only means of shortening the physical, equipment-based lead time. In shifting to a pull production system you will inevitably impact the information-based side of the production management role as well. The information flow within the factory changes completely and ultimately has a significant impact on the production schedule and planning process. See Figure 2-3.

Key Point

Lean production is a market-oriented system. In pull production the trigger for production is no longer the arrival of materials and parts based on a forecast of customer orders; it is the customer

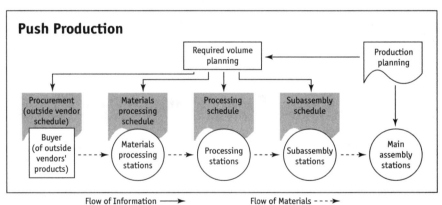

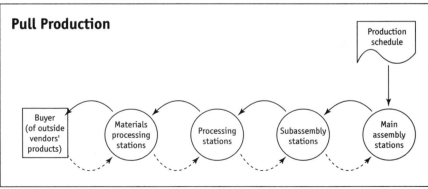

Figure 2-3. The Flow of Information and Materials in Push and Pull Production Systems

order itself. The critical shift to make at this point is from thinking of customer lead time as "transportation lead time from the warehouse" to thinking of customer lead time as "the whole process." See Figure 2-4.

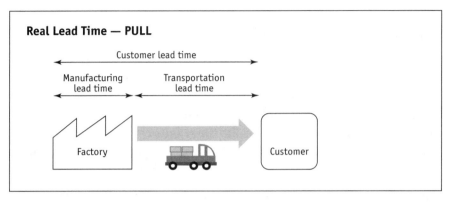
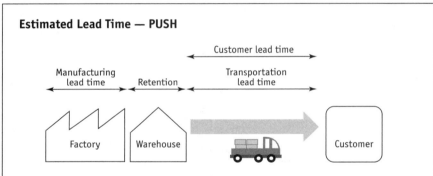

Figure 2-4. Two Views of Customer Lead Time

This shift in thinking will compel you to make the required improvements in the production process to shorten that time until it matches the customer's needs. By creating warehousing buffers and defining customer lead-time as transportation time from the warehouse, you lessen your ability to respond quickly to changes in the market, a serious flaw in today's economy.

Key Point

A strong, healthy factory is one that can meet needs for prompt delivery based on real lead time. This means attacking the most costly waste in a push system—inventory and overproduction. This is the purpose of implementing pull production.

The Problem with Stock and Overproduction

To understand how the pull system addresses the most costly waste in a production process we need to look at the problem of stock and overproduction.

Why Does Inventory Accumulate?

Finished good inventory accumulates as a result of the production strategy to create warehoused buffers to meet customer demand, as discussed above. As a consequence of this strategy inventory also accumulates throughout the production process. Just-in-time or lean production challenges the assumptions of this push production strategy and focuses on eliminating the causes of inventory build-up wherever it is found. You must understand why inventory accumulates and address these causes as soon as you uncover them. Inventory accumulates for many reasons, some of which are listed below.

- Shish-kabob production creates floods at certain processes
- Unbalanced capacity leads to unbalanced inventory
- Inventory is gathered from several processes
- Inventory must wait to be distributed from large processes
- Inventory must wait for a busy operator
- Operators have difficult changeovers
- People forget to revise standards after improvements are made
- Faulty production scheduling

- End-of the month rushes

- Seasonal adjustments

- Just-in-case inventory

Everyone wants a buffer for the unexpected. Safety margins are often maintained *just in case* a sudden change occurs in the production plan. These security measures *seem* to protect the production management process but actually they only serve to hide the problems that exist in the process. As we saw in Figure 2-3, shifting to pull production reverses the flow of production scheduling information in the factory by triggering production based on actual customer orders in hand. Through kanban and the other tools of pull production discussed in Chapter 4, downstream processes trigger upstream processes ultimately back to material procurement itself. When suppliers begin to conform to the standards of your pull system, factory flow based on customer orders will be a reality. Until then, improvements in the manufacturing process must continually be directed to eliminating the need for work-in-process inventories.

Why Is Inventory Bad?

In Japan, inventory is considered "the company's graveyard." It is considered the likely cause of poor performance in any business activity. Why is inventory so bad?

1. Inventory solidifies capital without turning it over to profit. See Figure 2-5.

2. Inventory takes up space and incurs maintenance costs—warehouse, insurance, taxes, etc.

3. Excess inventory leads to losses from obsolescence and subsequent price-cutting.

4. Inventory creates wasteful operations and energy consumption— conveyance, counting, picking up, putting down; electric, hydraulic, pneumatic equipment usage; and additional management to keep track of inventory status.

5. Advance procurement of parts and materials may not match actual orders.

6. Inventory conceals deep-rooted problems in the system. See Figure 2-6.

Balance Sheet Summary

Goods not yet sold

Assets		Liabilities	
1. Current assets		**1. Current liabilities**	
Cash	xx	Notes payable	xx
Notes receivable	xx	Accounts payable—trade	xx
Account receivable—trade	xx	Short-term loans payable	xx
Finished goods	xx	Accrued amount payable	xx
Work-in-progress	xx	and accrued expense	
Materials	xx	Allowance for taxes	xx
2. Fixed assets		**2. Fixed liabilities**	
Property, plant, equipment		Bonds payable	xx
Buildings	xx	Long-term loans payable	xx
Machinery and equipment	xx	Allowance for employee	xx
Land	xx	retirement benefits	
Intangible fixed assets		**Total Liabilities**	**xxx**
Good will	xx	**Shareholders' equity**	
Patent rights	xx	**1. Capital stock**	xx
Investments and other assets		**2. Legal reserve of retained earnings**	xx
Investments in securities	xx	**3. Surplus**	
Investments	xx	Voluntary reserves	xx
3. Deferred charges	xx	Unappropriated retained	xx
		earnings at end of term	
		(Earnings at end of term)	(xx)
		Total shareholders' equity	xx
Total Assets	**xxx**	**Total Liability and**	**xxx**
		Shareholders' Equity	

Investment dollars

Figure 2-5. Inventory Gives No Return on Investment Until It Has Been Sold

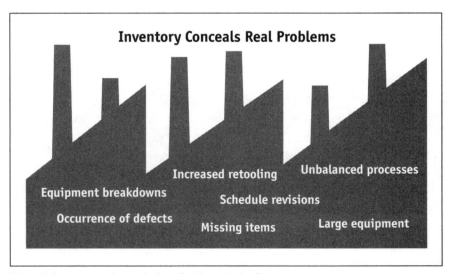

Figure 2-6. Inventory Conceals Real Problems in the Factory

One of the most difficult things to change in implementing a pull system is the mindset of people who depend on inventory. *Changing everyone's mind about inventory will be the most important accomplishment you make in preparing to switch to the pull method of production.*

Key Point

We're doing great. We're maintaining a month's worth of inventory.

Wait a minute. I thought we were supposed to have almost no inventory.

Figure 2-7. What Is the Appropriate Amount of Inventory?

TAKE FIVE

Take five minutes to think about these questions and to write down your answers:

1. Why does inventory accumulate?
2. What is the value of eliminating stockpiles of work-in-process inventory?
3. What is the alternative?
4. What concerns do you have about shifting to a pull system?

Push versus Pull Production

The original term for the Toyota Production System was just-in-time (JIT) manufacturing, which highlights the goal of TPS to manufacture and procure "just what is needed, just when it is needed, and just in the amount needed." This applies to all aspects of the manufacturing process including procurement, subcontracting, and distribution. The more recent term, lean production, highlights the focus on removing waste from the production process. Whatever it is called, the pull production system delivers both of these goals. By reversing the order information process so that downstream operations pull material from upstream operations, starting with the customer order as the trigger for production, "just in time" becomes a reality. In doing so, waste is revealed that was hidden by the push process. When the waste is removed by reducing changeover times, eliminating WIP, improving layout, and creating flow, "lean" becomes a reality. Three unique aspects of pull production are discussed below.

Production Is Music

Music consists of melody, rhythm, and harmony. The *melody* in sheet music is a string of notes; in the factory *the melody is the flow of work pieces down the line*. Kanban and one-piece flow create a most melodic result—waiting, conveyance, inventory build-up or shortages, and defects have been eliminated. Music can be expressed through many *rhythms*—waltz, salsa, fox trot. In the factory *the rhythm is the pitch of production*, the rate that work-in-process flows through the production line. It is also the takt time or cycle time. Different products will flow at different rhythms. The rhythm must be "leveled" to keep in time with cycle time or halting and uneven rhythms result. *Harmony* comes from a pleasing blending of tones in music. In a factory, *when people,*

Figure 2-8. The Hum of Pull Production in the Lean Factory

machines, and materials work together according to waste-free standards and without hindrances, harmony results. When pull production functions the factory hums.

The Next Process Is Your Customer

At the beginning of this chapter we discussed the difference between push and pull production systems in terms of product-out versus market-in systems. What this means in terms of the actual production process is that in the pull system the downstream process is the customer of the upstream process. This differs completely from the push system where work pieces are sent to the next process as soon as they are completed, whether or not the downstream process is ready for them. In pull production, operators do not produce until they are signaled by the next process to do so. This functions like a super market where customers go to the shelves and withdraw what they need; and the shelves are replenished as items are withdrawn. In the pull system, operators of the downstream process withdraw a finished work piece from the previous process, "paying" for it with a kanban. The kanban serves as the production order for the upstream process to replace the piece that was withdrawn. (Kanban will be explained in more detail in Chapter 4.) The art of pull production is to maintain the minimum number of pieces ready for withdrawal at each process so that overproduction never occurs and the need for stockpiles of inventory is eliminated. See Figure 2-9.

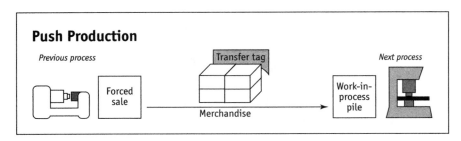

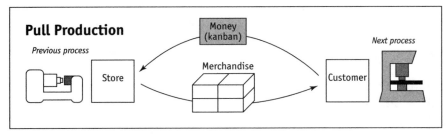

Figure 2-9. The Next Process Is Your Customer In Pull Production

Stop and Go versus Process and Go Production

There are four primary categories in the flow of materials through the production line:

1. Retention—waiting between processing points or transfer points, warehouse time

2. Transfer—moving materials from one retention point to another

3. Processing—the value-added activities

4. Inspection—defects are removed

In our music analogy, these four elements make a four-beat bar, which must be repeated continually. Because retention is typically such a large part of the process, this type of production "music" is called "stop-and-go production." In the example shown in Figure 2-10, there are eight retention stages or beats from the parts warehouse to the finished products warehouse. Six retention stages are within the factory. Retention time eats up a lot of lead-time; it is the worst offender in this regard. Transfer time takes usually only a few minutes unless you include time to transfer materials from outside vendors; it is the second worst culprit in consumption of lead-time.

Inspection can be considered an entirely different sort of lead-time culprit. Pull production internalizes the inspection process so that no defective products move past the workstation where the defect occurs. The downstream process serves as back up, empowered to reject and return any defect that may have slipped through. Error-proofing devices, andon lights, and workers empowered to stop the machines when any defect occurs (jidoka) eliminate the need for a separate inspection function.

Finally there is the "processing" element, which eats up the least amount of lead-time and is the only part of the process that is value-added activity. Most individual processes take less than a minute, many only a few seconds. Bringing in new production equipment to shorten process time is not the best way to reduce lead-time. Thus, in lean production, the equipment is the last thing to be addressed. First eliminate the wasteful aspects of lead-time activities.

Key Point

The best way to shorten manufacturing lead time is to eliminate retention (Figure 2-10). Once this is done, "stop-and-go" production with a four-beat bar of process-retention-process-retention becomes "process-and-go" production with a four-beat bar of process-transfer-process-transfer. The production hum begins to be heard.

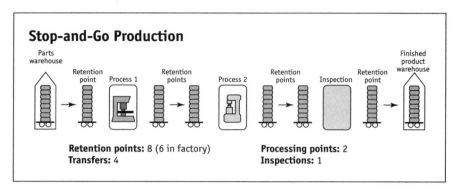

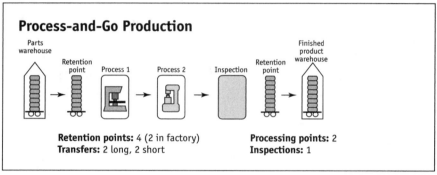

Figure 2-10. Stop-and-Go versus Process-and-Go Production

What Is Flow Production?

Key Term

Flow is *the progressive achievement of tasks in the creation of a product without stoppages, scrap, or backflows.* Flow can apply to the transmission of information as well as the movement of materials.

The Value Stream

Another way to think about flow is in terms of the "value stream." The value stream is all the activities in your company that are needed to design and manufacture product and deliver it to your customer—all the lead-time activities discussed throughout this chapter. "Value" is the worth of a product or service delivered to a customer. It is the degree to which a customer need or desire is fulfilled and may include quality, usefulness, functionality, availability, price, beauty, and so on. "Value-added" refers, then, to any operation in the value stream that changes raw material into value for the customer. Eliminating waste in the value stream seeks to reduce the non-value-added activities so that the value stream consists predominantly of value-added operations. *When the value stream consists primarily of value-added activities then flow will result. In the Toyota Production System, pull production serves to deliver this result.*

Key Point

Seven Requirements for Flow

There are seven requirements for achieving flow:

1. Arrange the production processes and machines in the sequence of their process—line or U-shaped cells.

2. Install smaller, slower, and more specialized equipment, keeping general-purpose equipment to facilitate flexible reorganization where needed.

3. Establish one-piece flow.

4. Synchronize processes to keep pace with client needs and the needs of the next process.

5. Use multi-process handling so that one worker can move from process to process down the line, sometimes handling an entire U-shaped cell alone.

6. Train workers in the multiple skills they will need for multi-process handling.

7. Change work position from sitting to standing so that workers have mobility to move through several operations if needed.

TAKE FIVE

Take five minutes to think about these questions and to write down your answers:

1. What are a few ways in which pull production differs from push production?

2. How is the production process like music?

3. What is the value stream?

Leveling

Key Terms

Leveling means to *completely even out the production of product types and volumes in relation to customer orders and customer needs.* Break down the monthly production output into daily amounts. Compare the daily volume of products with the operating hours and calculate how many minutes it will take to make each unit. This unit production time is called "cycle time" or "takt time." Then perform line balancing—*figure out how many people are needed to meet the production demand.* In the pull system, production is based on how many and what types of products are actually ordered, not how many products the factory is capable of making in a day. By removing the accumulated inventory of products that need to be shipped and sold, you are better able to design the latest customer needs into the factory process. Chapter 3 discusses leveling and line balancing in more detail.

Implementation Sequence

Key Point

As you prepare to implement pull production, plan to begin with the processes closest to the external customer—product inventory and final assembly. Then gradually move upstream with your implementation. Once the assembly and subassembly lines have been converted you can move laterally to lines for other models. Then move vertically again, shifting the processing lines, materials processing lines, and finally procurement. See Figure 2-11. The next chapter discusses the critical issues involved in implementing pull production.

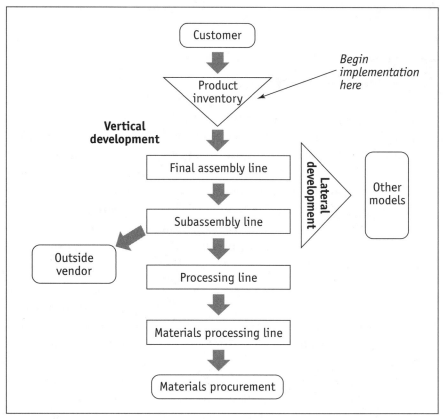

Figure 2-11. The Sequence for Introducing Pull Production

In Conclusion

SUMMARY

The increase in customer demand for wider variety in products and product features requires that manufacturers produce in small lots while still maintaining short delivery schedules and high quality. The old QCD formula is no longer sufficient; diversity in products must be added to quality, cost, and delivery excellence. Manufacturing companies will not survive without the ability to manage wide-variety, small-lot pull production. The ability to deliver greater diversity while keeping costs down requires lean production methods and, in particular, a pull production system.

Factories are living organisms with both information-based and equipment-based mechanisms that support the flow of goods to the customer. Production management consists of these two important aspects: the information-based management or "paper" system, and the equipment-based or "physical" system. Efforts to eliminate waste must address both aspects of factory life. The information-based aspect of production management is often addressed first—with computerization. But if the factory organization remains locked into the rules of large-lot mass production, these computer systems will be at odds with any changes you make in the equipment-based aspect of production management toward wide-variety, small-lot production. *Once pull production is established much of the information-based process will be eliminated or dramatically altered.* As you move forward in your implementation of pull production you will need to examine these systems carefully. Hopefully you will not confront too much resistance to the changes you will want to make on the "paper" side of your organization.

If we examine the issue of product lead-time, the two aspects of production management come into focus. *Product lead time begins with the production plan and ends with product shipment.* There is much more to the paper side than the physical side of a production plan. Only when production orders are issued do materials begin to flow. The factory-based improvements addressed by lean production methods are the only means of shortening the physical, equipment-based lead-time. In shifting to a pull production system you will inevitably impact the information-based side of the production management role as well. The information flow within the factory changes completely and ultimately has a significant impact on the production schedule and planning process.

Lean production is a market-oriented system. In pull production the trigger for production is no longer the arrival of materials and parts based on a forecast of customer orders; it is the customer order itself. *The critical shift to make at this point is from thinking of customer lead time as 'transportation lead time from the warehouse" to thinking of customer lead time as "the whole process."* This shift in thinking will compel you to make the required improvements in the production process to shorten that time until it matches the customer's needs. A *strong, healthy factory is one that can meet needs for prompt delivery based on real lead time. This means attacking the most costly waste in a push system—inventory and overproduction.* This is the purpose of implementing pull production.

Inventory accumulates for many reasons. Everyone wants a buffer for the unexpected. Safety margins are often maintained *just in case* a sudden change occurs in the production plan. These security measures *seem* to protect the production management process but actually they only serve to hide the problems that exist in the process. In Japan, inventory is considered "the company's graveyard." It is considered the likely cause of poor performance in any business activity. It solidifies capital without turning it over to profit and it incurs maintenance costs—warehouse, insurance, taxes, etc. Excess inventory leads to losses from obsolescence and subsequent price-cutting. It creates wasteful operations and energy consumption. Advance procurement of parts and materials may not match actual orders. But most importantly, inventory conceals deep-rooted problems in the system. The art of pull production is to maintain the minimum number of pieces ready for withdrawal at each process so that overproduction never occurs and the need for stockpiles of inventory is eliminated. *Changing everyone's mind about inventory will be the most important accomplishment you make in preparing to switch to the pull method of production.*

Flow is the progressive achievement of tasks in the creation of a product without stoppages, scrap, or backflows. Flow can apply to the transmission of information as well as the movement of materials. Another way to think about flow is in terms of the "value stream." The value stream is all the activities in your company that are needed to design and produce product and deliver it to your customer. *When the value stream consists primarily of value-added activities then flow will result. In the Toyota Production System, pull production serves to deliver this result.*

As you prepare to implement pull production, plan to begin with the value stream processes closest to the external customer—product inventory and final assembly. Then gradually move upstream with your implementation.

REFLECTIONS

Now that you have completed this chapter, take five minutes to think about these questions and to write down your answers:

- What did you learn from reading this chapter that stands out as particularly useful or interesting?

- Do you have any questions about the topics presented in this chapter? If so, what are they?

- What additional information do you need to fully understand the ideas presented in this chapter?

Chapter 3

Implementing Pull Production

The Improvement Infrastructure

The prerequisites for implementing pull production were discussed in Chapter 1. In Figure 1-4, an awareness revolution and 5S were shown as the foundation steps for implementing lean production. This means that the entire company understands the importance of eliminating waste in their operations and processes and is involved in team-based improvement activities. It also means that a lean promotional organization has evolved to support ongoing improvement activities, to train the workforce in new methodologies, and to continually revitalize the lean focus. Figure 3-1 is a good example of the people who need to be involved in the lean production infrastructure. Notice that managers and supervisors from the top of the organization to the factory floor are directly and actively immersed in the improvement system.

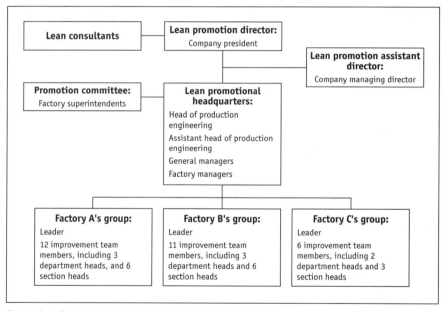

Figure 3-1. The Lean Production Improvement Infrastructure

Key Roles

If such an organization does not yet exist, take the time, *now*, before beginning the pull implementation, to identify the key roles needed to fill out this organization. You will probably find that many people are already functioning in these roles. Identifying these people, making their lean roles part of their formal job descriptions, and announcing the structure of the lean promotion organization will send a message throughout the company that lean is a serious effort. Most importantly, doing this will make it much easier to carry out the complex implementation of pull production that lies ahead.

The Promotion Office

With this organization in place, if you have not yet done so, create a lean promotion office in the factory where improvement team meetings can be held. This is not a recreation room but a team workroom.

Improvement Days

Key Point

Structure a weekly cycle of improvement days for each team. Ideally, you should have one whole day per week dedicated to improvement efforts. During the pull implementation, you may need to set aside a full week to make changes in plant layout for each cell. But generally, one day per week until the implementation is complete will make it possible to shift to pull without unduly affecting the production schedule. The habit of once-a-week improvement days may already exist in your plant if you have been implanting 5S. Some plants start with half days, or identify a number of hours each individual is to spend involved in improvement activities. This makes it possible to stagger the time associates are involved in activities other than production operations so that production is not interrupted.

Improvement Meetings

Teams benefit from meeting once a week to share progress reports, brainstorm, and assign improvement tasks. When engaged in the implementation of specific methods, as in pull production, the habit of weekly improvement meetings will support success.

Key Point

Meetings are not the goal of lean production, improvement is. So meetings should be short and to the point. One hour is better than two! But don't hold meetings at mealtimes or scheduled breaks. Improvement must be recognized as part of people's jobs, which they are being paid to carefully carry out. This will be crucial as you begin to shift to pull production. Having a meeting at the beginning of an improvement day is a good plan. Teams can share the results of the previous week's improvement efforts and set targets for the rest of the day.

An Improvement List

Create an improvement list. If you walk through the factory you can probably identify ten to fifteen areas in need of improvement. These can be the focus for ongoing team improvement activities. This list should include columns to write in the improvement number, name, description, department or section, leader, delivery deadline, and result.

When implementing the steps in the pull system, the improvement list will consist of the activities involved in the implementation process itself. It will help focus people's efforts in the right direction.

The Factory Floor

Key Point

The majority of improvement time should be spent making improvements on the factory floor. Allocating special time for this creates focus and speeds progress. When you set aside a special day for improvement, teams can concentrate on actually making improvements. Changes that are made on that day can be monitored and refined throughout the rest of the week. Reports on the success of the improvements and ways to solve problems that have occurred can be discussed at the next week's meeting. You will have some order to the process. If teams have to scavenge for time throughout the week to make improvements, the process of improvement will go more slowly. Each change will take longer to be refined and discussed.

Kaizen Events

Another approach to making major changes in the factory is to hold a kaizen event. This is typically a weeklong, carefully planned implementation period where the target line is shut down, and a special team comprised of different experts throughout the plant and some outside specialists join to overhaul an entire line. This is often the approach taken when shifting to cell design—the first stage of moving to a pull system.

Some companies have adapted the weeklong event to shorter blitzes of half-, one-, two-, or three-day events. The kaizen event has specific phases and requires detailed planning to be successful. Refer to *Kaizen for the Shopfloor*, listed in the reference section at the back of this book, for information on how to conduct these events.

However you decide to allocate time for improvement activities, the key point here is that improvement must be considered part of people's assigned duties and time must be set aside for any implementation or sustained improvement to succeed.

TAKE FIVE

Take five minutes to think about these questions and to write down your answers:

1. Do you have an improvement office allocated on the factory floor?

2. How often do improvement teams meet?

3. Is the company president involved in your improvement program? What other company managers participate?

Five Steps to Pull Production

To implement pull production, plan to move through five steps of your production process before the implementation is complete.

How-to Steps

Step One: Identify the Current Process

You probably have a traditional plant layout where machines for each operation are placed together and operators are trained to work only on those machines for that one operation. This is called an "operation-based" layout. In Figure 3-2, for instance, the lathes are placed together, the milling machines are placed together, and so on; but no attention has been given to placing them in the sequence in which they are used to make the product. The first step is to identify the flow of material through the existing system. Where does the process begin? Where does it end?

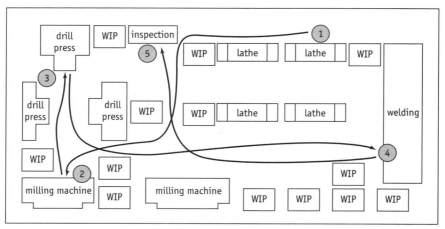

Figure 3-2. An Operation-Based Layout

TAKE FIVE

Take five minutes to think about these questions and to write down your answers:

1. Choose one production process. What is the total distance between each operation?

2. How much *time* does this distance represent?

How-to Steps

Step Two: Co-Locate Equipment in Sequence

The next step is to shift from an operation-based to a process-based layout by co-locating the equipment involved in a process sequence. See Figure 3-3. The transportation time between operations will be greatly shortened. Although WIP inventory will not yet be reduced, it will be easy for everyone to see its uselessness.

Key Term

Monuments. *Monuments* are *large machines that are difficult and expensive to move.* Where monuments exist the process can be organized around them so that they won't have to be moved. See Figure 3-4. If these immovable machines are shared by several processes or used in batch processing, cells can be designed around them. See Figure 3-5. Even better would be to switch to smaller more flexible machines.

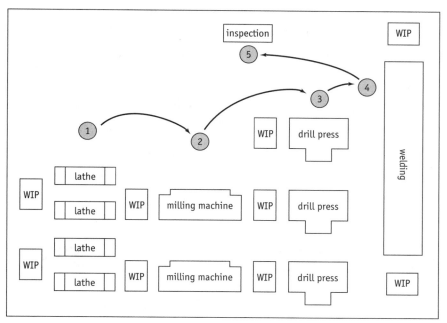

Figure 3-3. A Process-Based Layou

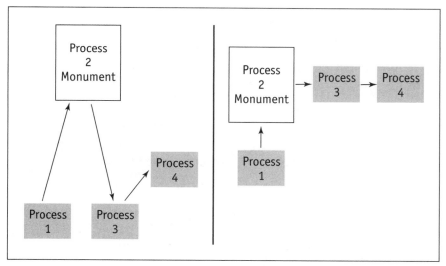

Figure 3-4. Move the Process to the Monument

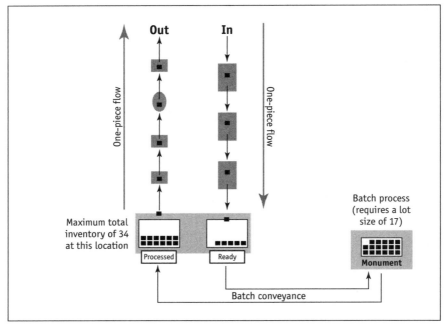

Figure 3-5. Cell Design Using a Monument for Controlled Batch Processing

TAKE FIVE

Take five minutes to think about these questions and to write down your answers:

1. How many processes do you have?
2. Do you have any monuments in your factory?
3. Do you need to plan kaizen events to change the layout so that equipment co-location can be accomplished quickly?

How-to Steps

Step Three: Design Manufacturing Cells

Once the equipment has been co-located according to process, the next step is to design manufacturing cells. See Figure 3-6. Some companies may be able to go directly from an operation-based layout to cells, without having to first co-locate equipment in sequence. It depends on the time you can devote to improvement and the complexity of your processes. In Figure 3-7, improvement is carried further as sequencing shifts from process-based to product-based, where each product or product family has its own line.

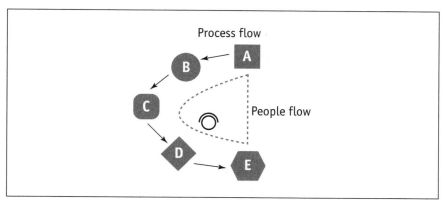

Figure 3-6. Cellular Manufacturing

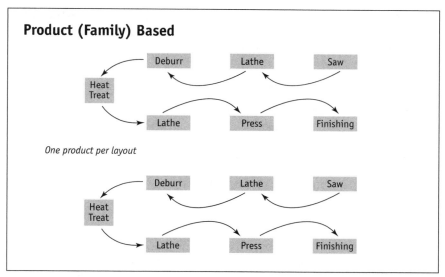

Figure 3-7. A Product-Based Layout

Sequence. Organize the sequence to reduce or eliminate the WIP inventory stacks and further reduce transport distances between operations. When you organize machines sequentially you create some challenges: Delays will result from unreliable equipment; equipment may be idle at times; and line balancing (discussed later in this chapter) will be required. However, the benefits far outweigh the challenges: eliminating the majority of transportation time will dramatically reduce lead-time; floor space will be freed up; and WIP inventories will be reduced to a minimum.

Key Point

Perhaps *the most interesting and valuable intangible benefit of cell design is increased operator awareness of the entire process.* An

43

operator's ability to see the process better allows him or her to think about it and improve it. Operators can work together more easily to solve problems; the idea of the downstream customer becomes a reality and pride in creating defect-free products is more personally embraced. Improvement ideas and operator involvement in the improvement process increases greatly once cell design is in place.

Cell shapes. Cell design is usually thought of as U-shaped, but cells can be designed in many shapes. The key is that they are process-based, not operation-based, as discussed earlier. Figure 3-8 shows a variety of cell shapes that can be used effectively. Cells can be in arranged in straight lines, L-shapes, U-shapes, in shapes like an = sign (or curtain), and S-shapes. Choose shapes that best suit the type of product being made; the materials or operations involved; the type, size, and shape of the equipment in the process; the way materials must be delivered to the cell or finished pieces withdrawn from it; or the relationship of the cell to the upstream and downstream processes it comes between.

The advantages of the U-shaped cell include the following:

1. Work enters and leaves at the same place.

2. Communication between people increases.

3. Walking distance is shortened between operations.

4. Multiple operations and line balancing become easier.

Point of use. *Each operation should be as close as possible to the next; this prepares the cell for one-piece flow.* The advantage of cell design is to eliminate delivery distances and time as well as turning, twisting, lifting, and reaching. Many improvements will have been made in this regard during your 5S activities. Once cell design has been achieved, operators will discover many additional ways to improve operations. *Point of use means that all needed supplies are within arms reach and are positioned in the sequence they are used.*

Straight line

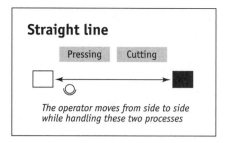

The operator moves from side to side while handling these two processes

L-shaped line

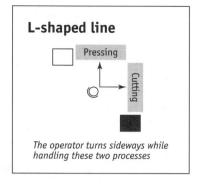

The operator turns sideways while handling these two processes

U-shaped cell

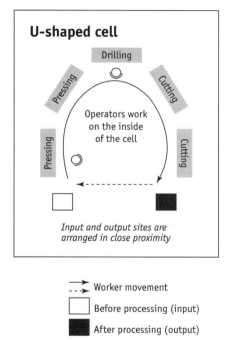

Operators work on the inside of the cell

Input and output sites are arranged in close proximity

→ Worker movement

--→ Before processing (input)

■ After processing (output)

S-shaped cell

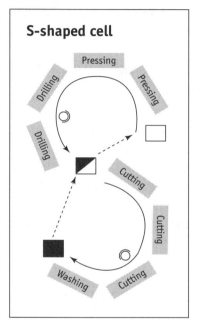

Cell shaped like an equal sign (=)

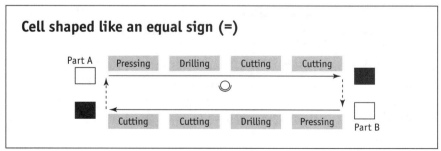

Figure 3-8. Cell Design Shapes

How-to Steps

Step Four: Initiate the Kanban System

At this point, the essence of pull production—kanban—can be initiated. The kanban system is also called the nervous system of the factory. Kanbans serve as the production order for the pull system. Order information in a pull system travels upstream from sales to assembly instead of downstream from planning and procurement. Kanbans follow the goods and indicate what is to be withdrawn from the upstream process. As soon as a client orders a product, a work order is sent to the assembly line, which in turn orders parts from the process lines. The process lines order the materials needed from procurement and so on. This is the reverse of the push system.

Implement the kanban system according to the sequence shown in the previous chapter in Figure 2-11, starting at the product inventory warehouse and final assembly. Once the system is stabilized there, then begin to address the subassembly processes. It will take time and patience to shift the entire factory to this system so that materials procurement occurs as part of the pull system. There is a great deal of information available about how to use MRPII with the kanban system to help you make this shift. Essentially you are focusing on eliminating the need for inventory at each stage of the process. The interim solutions will include a variety of methods, including a two-stage pull system using "supermarkets" between processes (discussed in the next chapter).

Chapter 4 will discuss the tools of the kanban system. *Kanban for the Shopfloor*, listed in the reference section at the back of this book, provides details about using these tools, as do other books listed there.

TAKE FIVE

Take five minutes to think about these questions and to write down your answers:

1. Why is kanban the reverse of the push system?

2. What is the best sequence for implementing kanban?

How-to Steps

Key Term

Step Five: Shift to One-Piece Flow

The final step in pull implementation is shifting to one-piece flow. *One-piece flow is the process of manufacturing products one unit at a time at a rate determined by the needs of the customer.* Many processes may already have made this shift. If not, consider what processes still need to function this way and do what is necessary to make the transition.

One-piece flow has many benefits. It:

- Delivers products to customers with less delay.

- Supports small-lot, wide-variety production.

- Reduces storage and transport requirements.

- Lowers the risk of damage, deterioration, or obsolescence.

- Exposes other problems in the process that need to be improved.

Chapter 5 will discuss the one-piece flow process in more detail. Other books that discuss one-piece flow, including *Cellular Manufacturing: One-Piece Flow for Workteams*, can be found in the reference section at the end of this book.

TAKE FIVE

Take five minutes to think about these questions and to write down your answers:

1. What is one-piece flow?
2. How does it affect lead-time?
3. Why is small-lot, wide-variety production an advantage?

Level Production

In Chapter 2 we discussed the problems of large-lot production in relation to why inventory accumulates. We also discussed the powerful impact of small-lot production in reducing inventory by eliminating non-value-added elements of the process. See the reasons for large lots in Figure 3-9. It is important to remember that reducing lengthy setup times from hours to minutes makes small-lot, level production feasible.

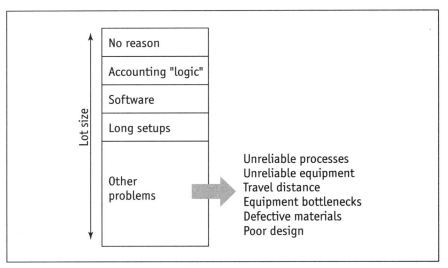

Figure 3-9. Why Are Lots Large?

Key Term

Now let's look at the process of shifting to small-lot production by leveling the production load. *Level production is a way of scheduling the daily production of different types of products in a sequence that evens out the peaks and valleys in the quantities produced.* It is also called load leveling, load smoothing, or heijunka. *Orders are sequenced in a repetitive pattern, smoothing day-to-day variations in total orders to correspond to longer-term demand.* See Figure 3-10.

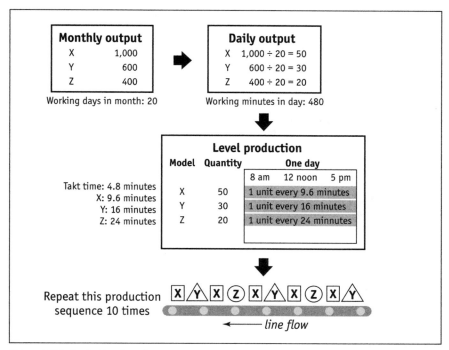

Figure 3-10. Production Leveling

Shish-Kabob Production

Mass production typically makes monthly schedules for each product, producing the entire quantity for each product in a sequence through the month. For example, 20,000 units of product A are produced in the first two weeks of the month, then 10,000 units of product B are produced, and finally 5,000 units of Product C at the end of the month. This type of production scheduling is called *shish-kabob production* because *product types move through the production process in large chunks like food on a skewer.* Figure 3-11 shows three types of scheduling, the first two are different forms of shish-kabob production. The first is mass, the second segmented. Only the third shows true mixed-model, level production.

Key Term

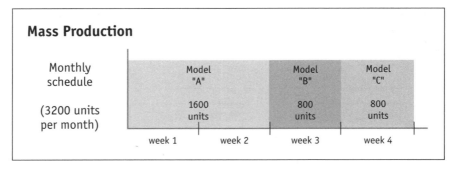

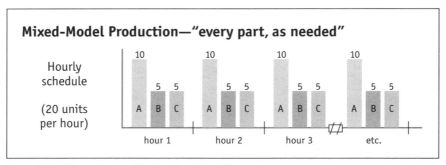

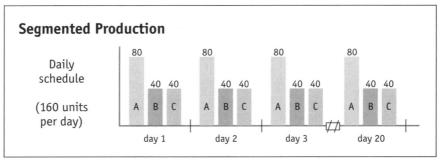

Figure 3-11. Three Types of Production Scheduling

Key Point

In level production, *the final process determines how many of each type of product must be made each day to meet customer requirements and, with mixed-model production, evenly mixes the required amounts in a smooth, repeating sequence.*

Takt Time

Key Term

The key to this daily schedule is a calculation called takt time. *Takt time is the rate at which products or parts must be produced to fulfill customer orders. It is not a measure of what you are capable of but a calculated number designed to match production to market demand.*

New Tool

The formula for calculating takt time is:

$$\text{Takt time} = \frac{\text{operating hours per day}}{\text{output demand per day}}$$

$$\text{Output demand per day} = \frac{\text{output demand per month}}{\text{operating days per month}}$$

See the example of calculating takt time in Figure 3-12. Takt time sets the rhythm or pace of production. In the music analogy discussed in Chapter 2, takt time would be the beat of the market. When orders are up, takt time sets a faster beat; when orders are down, takt time will be slower. Load leveling keeps the rhythm steady. Cell design, one-piece flow, and quick changeover are the prerequisites for realizing level production.

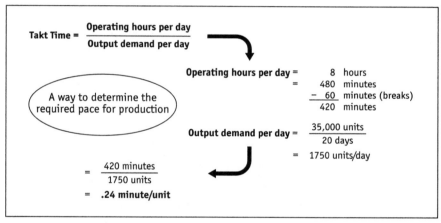

Figure 3-12. An Example of Takt Time Calculation

The Load Leveling Box

New Tool

A load leveling or scheduling box can be used to communicate to operators on each line the order of products to be produced throughout the day. See Figure 3-13.

Line Balancing

Key Term

Line balancing is a critical tool of pull production. *Line balancing is the process by which work is evenly distributed to workers to meet takt time.* Market demand determines the amount and types of product to be produced. As demand changes, lines can

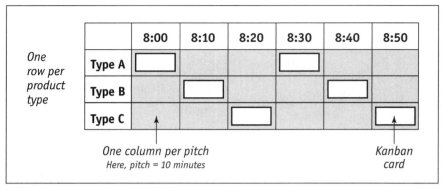

Figure 3-13. The Load Leveling Box

be balanced to match demand. When increased demand for a product exists, more people can be moved into the cells producing that product. When demand declines, fewer people are needed to operate the cell to meet the current need. Sometimes one person is enough to operate a cell. Other times someone will be needed at each operation in the cell. Use a *percent loading chart* like the one shown in Figure 3-14 to balance your operations.

New Tool

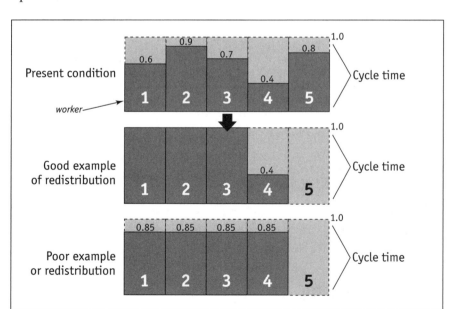

Figure 3-14. Line Balancing with Percent Loading Chart

Cross-Training

Key Point

Maximum flexibility requires cross-training of all operators. For pull implementation to succeed there must be union support for training programs that allow operators to learn to perform multiple functions. This can be on-the-job training and it makes each employee more valuable to his or her teams and to the company. It is also a source of pride for employees. Many companies display charts showing the skill attainment of employees. See Figure 3-15.

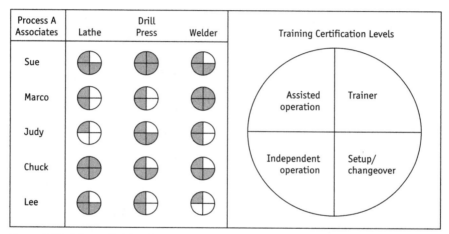

Figure 3-15. A Cross-Training Chart

Quality Systems

Key Point

In a pull system, quality inspection and prevention become every operator's responsibility.

Levels of Quality

There are five levels of quality systems in lean production:

1. Independent inspection

2. Operator inspection of products as they are made

3. Inspection by downstream operators as they are received

4. Mistake-proofing on errors; standardization

5. Supplier process control; suppliers never pass errors on

Operator inspection and inspection by downstream operators is achieved in pull production. The rules of quality in pull production are:

1. Never *allow* a defective product to be pulled *into* your workstation.

2. Never *produce* a defective product *at* your workstation.

3. Never *allow* a defective part to be pulled *from* your workstation.

Errors versus Defects

An error is some type of occurrence in the process that results in a defect. Improvement meetings can focus on solutions to errors and defects. The key to eliminating defects for good is to do root-cause analysis. This means separating errors from defects and searching for the causes of errors. By using a fish-bone diagram and the CEDAC process, errors can be identified, causes of errors can be brainstormed, and ideas for eliminating the causes can be sought.

TAKE FIVE

Take five minutes to think about these questions and to write down your answers:

1. What is level production?

2. What is the advantage of line balancing?

3. Is your union involved in cross-training programs?

4. What level of quality have you achieved in your manufacturing process?

In Conclusion

SUMMARY

It is important before implementing pull production that a lean promotional organization has evolved to support ongoing improvement activities, to train the workforce in new methodologies, and to continually revitalize the lean focus. If such an organization does not yet exist, take the time, *now*, before beginning the pull implementation, to identify the key roles needed to fill out this organization. You will probably find that many people are already functioning in these roles. Identifying these people, making their roles part of their formal job descriptions, and announcing the structure of the lean promotion organization will make it much easier to carry out the complex implementation of pull production that lies ahead.

You will need a lean promotion office in the factory where improvement team meetings can be held. This is not a recreation room but a team workroom. *Structure a weekly cycle of improvement days for each team.* Ideally, you should have one whole day per week dedicated to improvement efforts. *Meetings are not the goal of lean production, improvement is. So meetings should be short and to the point.* One hour is better than two! Having a meeting at the beginning of an improvement day is a good plan. Teams can share the results of the previous week's improvement efforts and set targets for the rest of the day. Create an improvement list. When implementing the steps in the pull system, the improvement list will consist of the activities involved in the implementation process itself. It will help focus people's efforts in the right direction. *The majority of improvement time should be spent making improvements on the factory floor.* Allocating special time for this creates focus and speeds progress.

There are five major steps to implementing pull production: (1) Identify the current process; (2) co-locate equipment in sequence; (3) design manufacturing cells; (4) initiate the kanban system; and (5) shift to one-piece flow. *Level production is a way of scheduling the daily production of different types of products in a sequence that evens out the peaks and valleys in the quantities produced. Orders are sequenced in a repetitive pattern, smoothing day-to-day variations in total orders to correspond to longer-term demand.* In level production, *the final process determines how many of each type of product must be made each day to meet customer requirements and evenly mixes the required amounts in a smooth, repeating sequence.*

Line balancing is a critical tool of pull production. *Line balancing is the process by which work is evenly distributed to workers to meet takt time.* Market demand determines the amount and types of product to be produced. As demand changes, lines can be balanced to match demand. When increased demand for a product exists, more people can be moved into the cells producing that product. When demand declines, fewer people are needed to operate the cell to meet the current need. Sometimes one person is enough to operate a cell. Other times someone will be needed at each operation in the cell. *Maximum flexibility requires cross-training of all operators.* For pull implementation to succeed there must be union support for training programs that allow operators to learn multiple functions. This can be on-the-job training and it makes each employee more valuable to his or her teams and to the company. It is also a source of pride for employees. Many companies display charts showing the skill attainment of employees.

In a pull system, quality inspection and prevention become every operator's responsibility. There are five levels of quality systems in lean production: independent inspection; operator inspection of products as they are made; inspection by downstream operators on products as they are received; mistake-proofing on errors—standardization; and supplier process control—suppliers never pass errors on. Operator inspection and inspection by downstream operators is achieved in pull production. The rules of quality in pull production are: never *allow* a defective product to be pulled *into* your workstation, never *produce* a defective product *at* your workstation, and never *allow* a defective part to be pulled *from* your workstation.

REFLECTIONS

Now that you have completed this chapter, take five minutes to think about these questions and to write down your answers:

- What did you learn from reading this chapter that stands out as particularly useful or interesting?

- Do you have any questions about the topics presented in this chapter? If so, what are they?

- What additional information do you need to fully understand the ideas presented in this chapter?

Chapter 4

Managing Pull Production

At this point the kanban system can be initiated. You have established an improvement infrastructure, identified your process, co-located equipment, and created production cells based not only on the process but the products being made. The rules of pull production and the kanban system are the way pull production is managed and synchronized throughout the plant to keep everyone responsive to the customer. In this chapter, we provide an overview of the kanban system. For more detailed information about using a kanban system see *Kanban for the Shopfloor*, listed in the reference section at the end of this book.

The Rules of Pull Production with Kanban

The rules of pull production define the significant differences between a pull and a push system. Every operator and supervisor must be fully aware of and trained in these rules and practice them.

Rule 1: Downstream processes withdraw items from upstream processes.

Rule 2: Upstream processes produce only what has been withdrawn.

Rule 3: All processes send only 100 percent defect-free products to the next process; no process accepts defects into their operations.

Rule 4: Level production is established to ensure market-in standards.

Rule 5: Kanbans always accompany the parts themselves for visual control.

Rule 6: The number of kanbans is decreased over time to reveal hidden areas for improvement.

Feedback with the Kanban System

As we have described in previous chapters, pull production works as a feedback system starting with the downstream customer. See Figure 4-1. The kanban system has emerged to coordinate production with the movement of parts and components between processes.

Key Term

Kanban means "card" or "sign" in Japanese. A kanban system uses cards and other visual signals to control the flow and production of materials. Let's examine the kanban system.

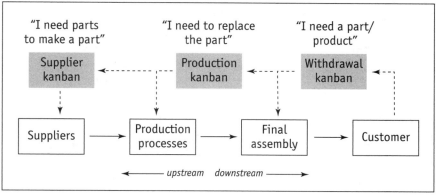

Figure 4-1. Feedback with the Kanban System

Functions of Kanban

The kanban system has several important functions.

A communication system. Kanbans are the communication system for lean production. They signal upstream processes when and what to produce and alert them when problems or changes occur so that production can stop. Kanbans signal standard operations to be engaged at any time based on actual conditions existing in the workplace. They also prevent unnecessary paperwork in start-up operations.

Pick up and work order information. Kanbans serve as work orders; they are automatic directional devices that provide two kinds of information:

1. What parts or products have been used and how many.

2. Where and how parts or products are to be produced.

The elimination of overproduction waste. Since production only occurs when signaled by a downstream process, in-process inventory and transportation are kept at a minimum and overproduction does not occur.

A tool for visual control. Since kanbans stay with the goods until the product is completed, they act as visual indicators of where production priorities exist and how operations are proceeding. Since they drive production, they are powerful visual controls of the process itself, determining when each process is to produce more and when it is to stop production.

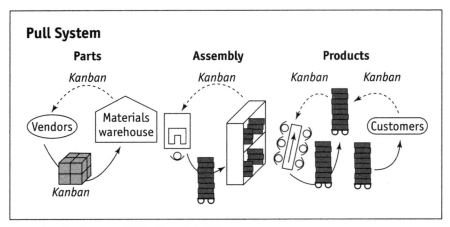

Figure 4-2. Kanban as a Production Work Order

A tool for promoting improvement. Inventory hides problems. Too many kanbans indicate excess in-process inventory. By reducing the number of kanbans, problem areas will come out of hiding so that they can be improved. In this way, the kanban system becomes a valuable means to drive out waste and continually improve the production system.

Reducing kanbans:

- Exposes problems in the order of their priority

- Reduces safety stock

- Stimulates improvement activities

- Allows you to fine tune your production cycles

- Let's you know how you are doing

- Helps you think about the right things

TAKE FIVE

Take five minutes to think about these questions and to write down your answers:

1. How does pull production work as a feedback system?
2. What are the functions of kanbans?
3. How does the kanban system help to improve factory operations?

Types of Kanban

Key Point

Kanbans are the "signs" attached to in-process inventory to indicate production orders. There are different kinds of kanbans.

Transport kanbans. The first major type of kanban is a transport kanban, which indicates when numerous parts are to be moved to the production line, or moved between processes in production and assembly. In addition to identifying the part and quantity, they indicate where the part comes from and where it is going. Transport kanbans include supplier kanbans and withdrawal kanbans.

Supplier kanbans, also called parts-ordering kanbans, are the orders given to outside suppliers for parts needed at assembly lines. If the kanban system has extended to the supplier network, then suppliers will deliver on demand as supplier kanbans are received from the factory.

Withdrawal kanbans, also called in-factory kanbans, are used between processes in the factory; they provide the details needed to withdraw parts from an upstream process. Withdrawal kanbans are used in many forms depending on the need and type of part being withdrawn—one kanban for a single part or for a container of parts, or a series of kanbans for when parts must be supplied in a certain order for downstream assembly sequences. These may also take the form of a kanban box, or a kanban cart for ease of transportation to the downstream process.

Production kanbans. The second major type of kanban is known as a production kanban. These indicate operation instructions for specific processes. Production kanbans include production-ordering kanbans and signal kanbans.

The first type is called a *production-ordering kanban*. These are the type that most people think of when they talk about the kanban system. They are routinely used at processes that do not require changeovers. A production-ordering kanban most closely resembles the standard production order used in a push system— it identifies what is to be produced and in what quantity. When a withdrawal kanban authorizes the removal of parts from a line or cell, a production-ordering kanban initiates production to replace the parts that have been removed.

Signal kanbans are used at presses or other processes that require changeovers. They signal when a changeover is needed in the sequence of production kanbans.

How the Kanban System Works

Each process has an inbound and an outbound stock area. The inbound area holds containers or pallets that contain a small, fixed quantity of materials, parts, or subassemblies used in the process. The outbound area holds the completed output from the process. Refer to Figure 4-3 as you read the following example of how the kanban system works.

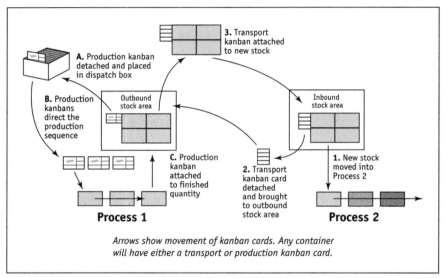

Figure 4-3. How a Kanban System Works

Each container in the inbound stock area has a *transport kanban* attached to it. When Process 2 begins to consume the contents of a container from its inbound stock area (step 1), the transport kanban is taken off the container (step 2) and brought to the outbound stock area of Process 1. There, the transport kanban is attached to a new, full container (step 3), which is taken back to the inbound stock area of Process 2, ready for use. A *production kanban* is attached to every container in the outbound stock area at Process 1. When Process 2 comes to remove a container of parts, the production kanban is taken off and put in a dispatch box for Process 1 (step A). Since Process 1 may make different parts for several other processes, it builds the new parts in the order in which the kanban are placed in the box (step B). When a container is filled with the required number of parts, the production kanban is re-attached to it (step C) and it is placed in the outbound stock area ready for withdrawal by Process 2.

How Many Kanban Do You Need?

The kanban system supports level production. It helps to maintain stable and efficient operations. The question of how many kanbans to use is a basic issue in running a kanban system. If your factory makes products using mostly standard, repeated operations, the number of kanbans can be determined using the formula in Figure 4-4.

$$\text{Number of kanban} = \frac{\text{Daily output} \times (\text{lead time} + \text{safety margin})}{\text{Pallet capacity}}$$

- Daily output = $\dfrac{\text{Monthly output}}{\text{Workdays in month}}$

- Lead-time = Manufacturing lead time (processing time + retention time) + lead time for kanban retrieval

- Safety margin: Zero days or as few days as possible

- Pallet capacity: Try to keep pallet contents small and instead increase the number of deliveries

Figure 4-4. How Many Kanban Do You Need?

Key Point

As you can see, *the number of kanbans you need is dependent on the number of pallets or containers and their capacity. Lead times, safety margins or buffer inventory, and transportation time for kanban retrieval are also important factors.*

Several questions must be answered when deciding the number of kanban to use:

1. How many products can be carried on a pallet?

2. How many transport lots are needed, given the frequency of transport?

3. Will a single product or mixed products be transported?

The relationship between the order-to-delivery period and the production cycle is also relevant. You will want to consider the following:

1. *Order-to-delivery time*. What quantity do assembly processes require and in what amount of time?

2. *Production cycle or lead time*. Consider:

   a. The time it takes to send withdrawal kanban to the previous process after removing the kanban.

   b. The time that elapses until processing is begun after exchanging withdrawal kanban for production kanban.

   c. The time it takes to produce supply lots.

   d. The time it takes to store the lot to be processed.

   e. The time it takes to transport processed items to the assembly line.

TAKE FIVE

Take five minutes to think about these questions and to write down your answers:

1. What are the two main types of kanbans?

2. Walk through the process of using the kanban system in one cell. Can you identify inbound and outbound stock areas?

3. Can you determine the lead time for the process including one cycle of withdrawal and production kanbans?

An Analysis of Pull Production

When Do You Pull?

Key Point

You pull material from the upstream process only as you need it to replenish what you have used to produce parts, components, or final assemblies. This is in direct contrast to the standard operating procedure of push systems: "If I have material at my station I will make more parts." Pull production requires behavior according to a new rule: "*I will not produce unless signaled!*"

Where Do You Pull From?

Upstream to you may be the preceding process or it may be a supplier of raw materials or subassemblies. You withdraw from the place where the materials are either produced or stored. Withdrawal kanbans identify the location of these materials so that there is no confusion about where to find them when you need them.

Places you might pull from include:

Examples

- A supplier
  a. An external supplier
  b. Other plants within your company

- An upstream operation

- A kanban quantity

- A feeder cell

- Supermarkets

What Do You Pull?

You withdraw only what you need to complete your operations. The material you pull from the upstream process can be contained in a number of ways. Whatever you pull must include a withdrawal kanban, which states the exact quantity that is contained in the pallet or lot that you withdraw. See Figure 4-5.

Figure 4-5. A Withdrawal Kanban Post

Examples

You may pull any of the following:

- A single part, a subassembly, or final assembly

- A container or pallet of parts

- A production lot

- A complete customer-order

Examples

Container options may include:

- Pallets
- Boxes of various sizes

- Bins
- Carts

- Trays
- Trucks, trailers, railroad cars

- Kits

What Do You Produce?

You only produce in order to replenish what has been withdrawn from your work process. This can be any of the following depending on your area or process.

- Raw material

- Parts

- Components or subassemblies

- Finished goods

If you work in the inventory warehouse for raw materials and parts purchased from outside suppliers or other plants, you will replenish raw materials as they are withdrawn for use in your subassembly production processes. If you work in subassembly you will produce the parts and components as they are withdrawn for final assembly. If you are in final assembly you will produce finished goods as customers order the products.

TAKE FIVE

Take five minutes to think about these questions and to write down your answers:

1. What is the new rule of behavior required by pull production?
2. Why is the new rule important?

One-Stage versus Two-Stage Pull Systems

One-stage pull production uses only one stock area or point of use for inbound materials and parts. See Figure 4-6. In a two-stage pull system, supermarkets are established between subprocesses to store kits for use by the downstream process. See Figure 4-7.

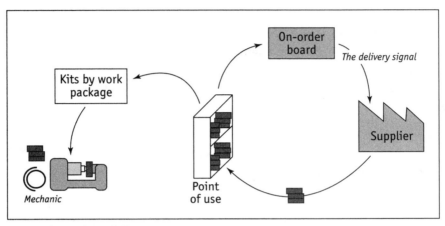

Figure 4-6. One-Stage Pull

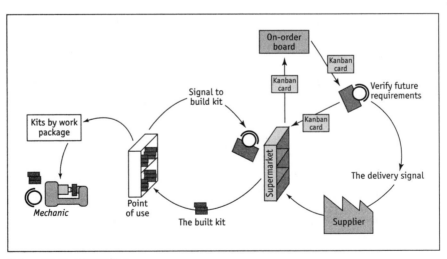

Figure 4-7. Two-Stage Pull

Supermarkets

Key Term

The kanban system was inspired by the way supermarkets work to supply customers with products. In the kanban system, *supermarkets* are *the areas where produced parts, components, subassemblies, and finished goods are stored.* These supermarkets or stores are located near the area where the parts are produced. When the downstream process (the customer) withdraws parts from an upstream process (the supermarket), the upstream process replenishes the withdrawn product by making another amount equal to what was withdrawn.

Withdrawal and production-ordering kanbans control this process, ensuring that there are always parts available when needed, but that no more than needed will be made. Supermarkets are most often used to store component kits between major steps in the process, reducing the number of places that downstream process workers have to go to withdraw the parts they need. The supermarket is the primary feature that distinguishes one-stage and two-stage pull systems.

Water Beetles and Milk Runs

Key Term

The term *water beetle* comes from the insect called a whirligig. It moves swiftly across the surface of water, spins, and makes rapid and unexpected changes in direction. In the Toyota Production System this name was given to *the conveyor—an operator who delivers parts to the other operators in his cell or on the line so that they don't have to get up to replenish their work stations to make the next set of products.*

In the kanban system, frequent transportation of parts becomes necessary. Usually hourly deliveries are needed. For simplicity, we have described the withdrawal and production process in terms of a single withdrawal location. The reality is that most production items require numerous parts that come from several preceding processes. This can be a time consuming and confusing aspect.

Key Point

Designating a carrier to withdraw parts from the preceding processes and deliver them to workers greatly simplifies the kanban process. Also, this person quickly becomes an expert in the withdrawal and production kanban process, making it possible to identify and eliminate errors.

Benefits of the water beetle. When a water beetle supports the line:

1. Operators don't have to leave the work area to find parts or tools.

2. Waiting time becomes visible.

3. The water beetle becomes the pace maker for the line; delays and progress become visible.

4. Water beetles can fill-in for absent operators or for less than "full work" operations.

Tools needed by the water beetle. To do the job well, the water beetle will need:

1. A pushcart with:
   a. Small wheels for easy maneuvering
   b. A checklist for pickups and deliveries to be made
   c. A layout of cells or lines so that pick up and placement can occur easily

2. A picking list to:
   a. Clearly identify *what* and where to pick up
   b. Clearly identify where and *when* to pick up
   c. Limit the number of items on the list

3. Visual displays to help identify supplies to be picked up or delivered

4. Pickup buckets for ease of transfer to and from the pushcart

5. Storage shelves where:
   a. Items are stored so that they are easily transferred to the pushcart—nothing above the shoulder or below the knee
   b. Gravity is used whenever possible to minimize lifting

6. A supply area at the line that:
   a. Makes transfer from pushcart to the line easy
   b. Is large enough for only one kit

Key Term

Milk runs. *Milk runs* are the name given to *the path the water beetle takes in his or her delivery and pickup runs.* Figure 4-8 shows a layout that allows the water beetle to serve a number of cells in one pass from supermarket and back again.

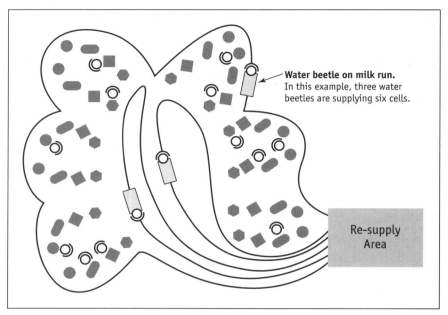

Water beetle on milk run.
In this example, three water beetles are supplying six cells.

Re-supply
Area

Figure 4-8. Milk Runs

TAKE FIVE

Take five minutes to think about these questions and to write down your answers:

1. Consider where a two-stage pull system might be beneficial in your factory.
2. What is the purpose of water beetles?

In Conclusion

SUMMARY

Pull production works as a feedback system starting with the downstream customer. The kanban system has emerged to coordinate production with the movement of parts and components between processes. *Kanban* means *"card" or "sign"* in Japanese. *A kanban system uses cards and other visual signals to control the flow and production of materials.* The kanban system functions as a communication system, as pick up and work order information, to eliminate overproduction waste, as a tool for visual control, and as a tool for promoting improvement. There are two major types of kanbans. Transport kanbans (including supplier and withdrawal kanbans) and and production kanbans (including production-ordering and signal kanbans).

The kanban system supports level production. It helps to maintain stable and efficient operations. The question of how many kanbans to use is a basic issue in running a kanban system. *The number of kanbans you need is dependent on the number of pallets or containers and their capacity. Lead times, safety margins or buffer inventory, and transportation time for kanban retrieval are also important factors.*

You pull material from the upstream process only as you need it to replenish what you have used to produce parts, components, or final assemblies. This is in direct contrast to the standard operating procedure of push systems: "If I have material at my station I will make more parts." Pull production requires behavior according to a new rule: "*I will not produce unless signaled!*"

Upstream to you may be the preceding process or it may be a supplier of raw materials or subassemblies. You withdraw from the place where the materials are either produced or stored. Withdrawal kanbans identify the location of these materials so that there is no confusion about where to find them when you need them.

You withdraw only what you need to complete your operations. The material you pull from the upstream process can be contained in a number of ways. Whatever you pull must include a withdrawal kanban, which states the exact quantity that is contained in the pallet or lot that you withdraw. You only produce in order to replenish what has been withdrawn from your work process. If you work in the inventory warehouse for raw materials and parts purchased from outside suppliers or other plants, you will replenish raw materials as they are withdrawn for use in your sub-assembly production processes. If you work in sub-assembly you will produce the parts and components as they are withdrawn for final assembly. If you are in final assembly you will produce finished goods as customers order the products.

One-stage pull production uses only one stock area or point of use for inbound materials and parts. In a two-stage pull system, supermarkets are established between subprocesses to store kits for use by the downstream process. The kanban system was inspired by the way supermarkets work to supply customers with products. In the kanban system, *supermarkets* are *the areas where produced parts, components, subassemblies, and finished goods are stored*. These supermarkets or stores are located near the area where products or parts are produced. When the downstream process (the customer) withdraws parts from an upstream process (the supermarket), the upstream process replenishes the withdrawn product by making another amount equal to what was withdrawn.

Withdrawal and production kanbans control this process, ensuring that there are always parts available when needed, but that no more than needed will be made. Supermarkets are most often used to store component kits between major steps in the process, reducing the number of places that downstream process workers have to go to withdraw the parts they need. The supermarket is the primary feature that distinguishes one-stage and two-stage pull systems.

A *water beetle is an operator who delivers parts to the other operators in his cell or on the line so that they don't have to get up to replenish their work stations to make the next set of products.* In the kanban system, frequent transportation of parts becomes necessary. For simplicity, we have described the withdrawal and production process in terms of a single withdrawal location. The reality is that most production items require numerous parts that come from several preceding processes. *Designating a carrier to withdraw parts from the preceding processes and deliver them to workers greatly simplifies the kanban process.* Also, this person quickly becomes an expert in the withdrawal and production kanbans, making it possible to identify and eliminate errors in the process. *Milk runs* are the name given to *the path the water beetle takes in his or her delivery and pick up runs.* Milk runs allow the water beetle to serve a number of cells in one pass from supermarket and back again.

REFLECTIONS

Now that you have completed this chapter, take five minutes to think about these questions and to write down your answers:

- What did you learn from reading this chapter that stands out as particularly useful or interesting?

- Do you have any questions about the topics presented in this chapter? If so, what are they?

- What additional information do you need to fully understand the ideas presented in this chapter?

Chapter 5

Extending Pull Production

Pull production, in theory, causes the factory to function as a single unit, pulsing to the beat of customer orders. In practice, this will increasingly become the reality as you install the kanban system and one-piece flow. One-piece flow is an ideal state that is not always possible, and, perhaps, not always desirable. The important thing is to promote a continuous flow of products with the least amount of delay or waiting in the process.

We have said throughout this book and all the books in this Shopfloor Series that 5S and quick changeover are prerequisites to pull production. At no time does this become more obvious than in shifting to one-piece flow. If you thought you had uncovered all hindrances, all waste, all retention points and causes of delay in your processes, *shift to one-piece flow and see what happens. You will be amazed at what instantly rises to the surface that you never realized might be causing problems for your operators. This is where the rubber meets the road.*

Key Point

Establishing One-Piece Flow Production

Key Point

There are several ways to shift to one-piece flow: cell by cell; product line by product line; or process by process, going from customer back to final assembly, final assembly to subassembly processes, subassembly to feeding processes, and finally to materials procurement. Some companies stabilize in one-piece flow at the point of customer order and final assembly. They have "supermarkets" of subassemblies close to the order point and final assembly is done order by order, assembling the product to suit each customer's unique requirements. This will certainly satisfy your customers, and is a great first step. However, until you tackle the subassembly and upstream, parts-production cells, you will not realize the great advantages of one-piece flow.

One-piece flow causes the total uncovering of concealed waste. Some recommend shifting to one-piece flow first, before co-locating equipment and designing cells, and even before implementing quick changeover. This is certainly a good way to uncover process waste. But depending on the type of product and the number of product lines you have, you will probably also have trouble keeping up with orders because of how much waste you encounter in your processes. Therefore, we suggest going in the order prescribed in this book, with one-piece flow as the *last step* in your pull implementation. Even doing it this way, plenty of hidden waste will be revealed.

In particular, you may discover the problem with those monuments you decided to hold on to. This may be the time to reconsider and invest in small, versatile equipment. Also, do not be afraid to move the equipment again and again. This is a work in progress. As you move to one-piece flow you will discover better ways to arrange things. As kanban kicks in it will become clear how to position cells in relation to each other so that further transport distances can be removed. Whether you move the equipment before or after implementing one-piece flow, most companies find that they have to rearrange the equipment a number of times before they get to an ideal state. It is a good idea is to install casters on the equipment.

Key Point

The benefit of having operators stand while working becomes evident now as well. Operators who resisted this before now may understand the benefit to themselves: *easier movement, helping each other out when necessary, quick correction of unbalanced operations, multi-process operations, and so on.*

TAKE FIVE

Take five minutes to think about these questions and to write down your answers:

1. What obstacles to one-piece flow remain in your production processes?
2. What area would make a good model line?

Seven Conditions for Flow

Remember the conditions for flow mentioned in Chapter 2? Let's review them:

1. Arrange the production processes and machines in the sequence of their process—line or U-shaped cells.

2. Install smaller, slower, and more specialized equipment, keeping general-purpose equipment to facilitate flexible reorganization where needed.

3. Establish one-piece flow.

4. Synchronize processes to keep pace with client needs and the needs of the next process.

5. Use multi-process handling so that one worker can move from process to process down the line, sometimes handling an entire U-shaped cell alone.

6. Train workers in the multiple skills they will need for multi-process handling.

7. Change work position from sitting to standing so that workers have mobility to move through several operations if needed.

You will run into many obstacles as you work to establish the conditions for flow in your factory, especially resistance in people's minds (including your own) about how this will work. *We recommend starting with the most enthusiastic workshop in your factory and creating a model line.* The model will show everyone in the company how flow production works in a line and what kinds of things it involves. It will give you a chance to work out the kinks in your own understanding of how to implement it as well.

Key Point

Steps to Achieve One-Piece Flow

Figure 5-1 shows the interrelationship of these conditions as steps to achieve flow. Note that steps 2 and 4 relate to the equipment, steps 3 and 5 relate to worker training, and steps 1 and 6 relate to the flow itself. All of these conditions have been discussed throughout this book, but a special note about number 6, synchronization, is important to mention here.

Key Term

Synchronization is when both processes and operators are synchronized to each other so that the entire line (and ultimately the entire production system) can become synchronized to market demand. This is accomplished with the pull production tools of level production and line balancing discussed in Chapter 3 and the kanban system discussed in Chapter 4.

When all conditions are in place, the ideal of one-piece flow can become a reality. With kanban, line balancing, and level production, your operations will be aligned with the customer and highly responsive to changes in the market.

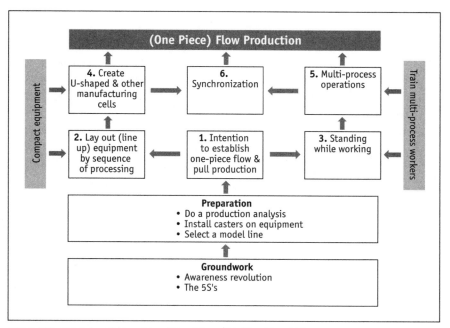

Figure 5-1. The Interrelationship of Conditions for (One-Piece) Flow

---

TAKE FIVE

Take five minutes to think about these questions and to write down your answers:

1. What conditions for flow do you still need to achieve?

2. What are the obstacles to doing so?

Using Moving Lines and Feeder Cells

In shifting to a process-sequenced and product-based cellular lay-out, you probably abandoned the long conveyors that you may have used to transport materials long distances from the warehouse and between functions in the old operation-based layout. We have not discussed the use of moving lines for cell-based processes, but these can be used effectively for some product lines. They should not be installed until you have eliminated the waste in your operations in each cell and the cell is functioning smoothly. At this point, a moving line can be used to help synchronize the timing of the cell to takt time—if you design it so that speed can be changed as demand changes, and so that linked operations can be handled by varying the number of operators in response to demand.

Under certain conditions, a moving line can make delivery of parts and materials at the point of use easier. Some of your cells will not be able to eliminate certain non-value-added efforts without a moving line to serve this purpose. A process that requires lifting heavy loads from one machine to the next should make use of a moving line. Another example of using a moving line would be small-piece work with multiple but very brief operations. The moving line would make it unnecessary for operators to have to look up to pass the finished piece to the next operator. The operator could simply place it back on the moving line. This would fail to be an advantage if an operator is working the cell alone, unless all the operations can be done in place. He or she might prefer to walk the piece through the process as one-piece flow; and the moving line would inhibit this option. See Figure 5-2.

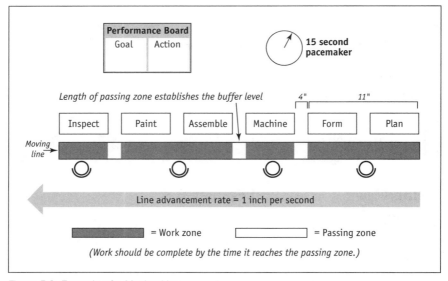

Figure 5-2. Example of a Moving Line

If you decide to use moving lines for some of your cells, *feeder cells* may be needed. Feeder cells can be used to deliver sub-assemblies to point of use effectively, eliminating the transport function. This is especially useful when individual operations are measured in seconds. See feeder cells in Figure 5-3.

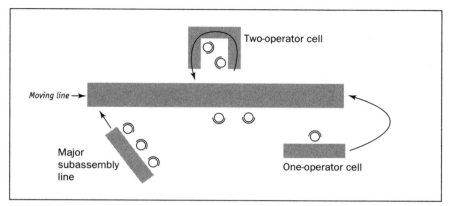

Figure 5-3. A Moving Line with Feeder Cells

TAKE FIVE

Take five minutes to think about these questions and to write down your answers:

1. Do you have any cells that would be enhanced by moving lines? Why?
2. Where might you use feeder cells?

Linking Suppliers to the Pull System

Once you have a successful model line of flow production, you can extend that clearly visible example to the rest of the lines in the factory. This is called lateral development. When flow production is well established throughout the plant you can begin to extend it to your vendors and subcontracted suppliers. This is called vertical development. *With suppliers, the aim is to reduce the amount delivered, and instead increase the number of deliveries.*

Changes in Transport Methods

To keep inventories down and lead times short, it will be necessary to significantly increase the number of smaller deliveries. Therefore delivery methods must be changed. If existing delivery methods are used, this will be too costly. Innovation in delivery will continue to be needed from the transport industry in three areas: loading methods for increased product diversification, frequency of deliveries for lower inventory levels and

Figure 5-4. Extending Pull to Suppliers

shorter lead times, and route planning for cost reduction. Below are some ideas that are currently being used by some transporters.

X

Examples

- *Loading methods:* Mixed products are loaded in the sequence of delivery to the receiving factory. See Figure 5-5.

Effects of the product diversification trend

Figure 5-5. Loading Methods

- *Frequency of deliveries:* The number of deliveries is increased to as many as 8, 16, or 32 a day instead of 1, and the amount delivered is decreased so that inventories and lead times at the receiving factory can be kept at a minimum. See Figure 5-6.

- *Route planning:* Instead of point-to-point deliveries, circuit or compound deliveries may prove to be more economical. See Figure 5-7 on page 86.

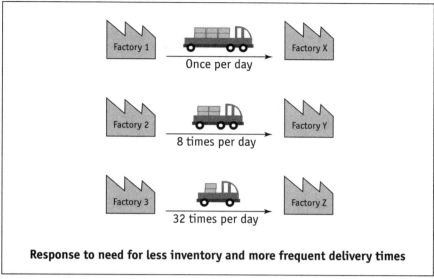

Response to need for less inventory and more frequent delivery times

Figure 5-6. Frequency of Deliveries

Changes in Delivery Sites

Once transport methods meet the needs of pull production, changes in where materials are delivered should be considered. Which part of the factory should take in the delivered items? Where and how deliveries are received can improve the ease and speed of handling materials within the plant and simplify the process of getting them to point of use on the line. *The closer deliveries can be made to point of use the better.* Consider the examples below to help eliminate waste in the flow of goods from the delivery sites.

- *Signboards:* Use signboards at the delivery site to identify who brings what to where and exactly when. See Figure 5-8 on page 87.

- *Color coding:* Select a different color for each line. Use the same color to identify materials going to that line and for the signboards for those materials at the delivery site.

- *Product-specific delivery sites:* If parts and materials are sorted according to the product they will be used for, instead of by the type or function of the part, waste will be eliminated as well.

- *FIFO—first in, first out:* Store items at the delivery site and at point of use so that the first item in is the first item used. Typically the opposite is done which causes products first in to be last used. Often the delay causes obsolescence. Reverse the storage method so that FIFO can be achieved. Circular conveyor belts can be used.

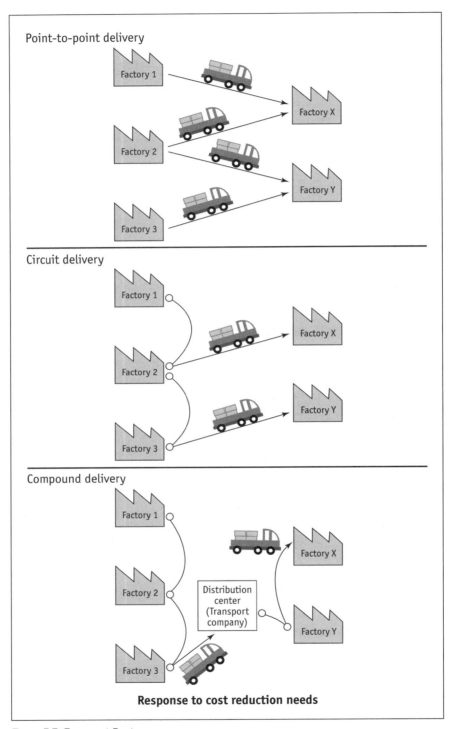

Figure 5-7. Transport Routes

• *Visible container organization:* Identify containers so that they are clearly distinguishable from each other, and so that parts within the containers are easily identified. See Figure 5-9.

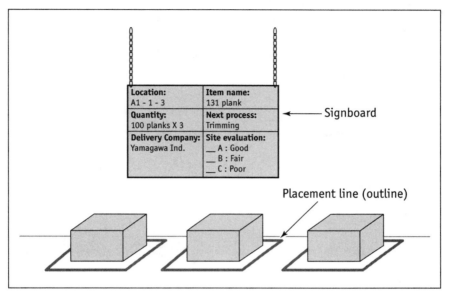

Figure 5-8. Signboards for Delivery Site Management

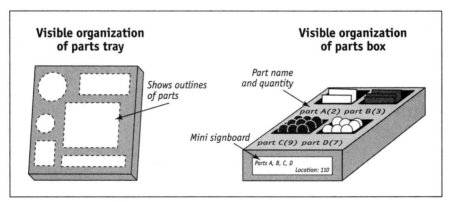

Figure 5-9. Visible Organization of Containers

Supplier Kanban

Electronic systems are increasingly replacing hand-delivered paper documents to manage procurement, but whether paper, fax, or e-mail, the supplier or purchasing kanban is used to make and track orders. It can also replace delivery vouchers. Figure 5-10 shows how the kanban system can be used with external suppliers who deliver parts and materials to the factory. This is an example of keeping the kanbans inside the factory. The kanbans for the supplies to be ordered are numbered. An order is placed identifying each item with its kanban number. When supplies are received they are checked against the kanban number thus replacing the delivery voucher. Everyone, including the transporter, knows where each item of the order goes. Delivery sites can be indicated on the supplier kanban, as well as the color code indicating the destination line or point of use.

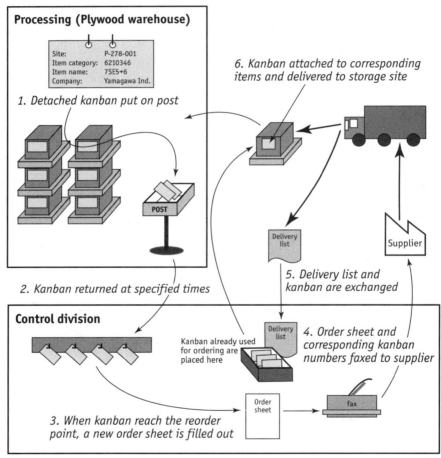

Figure 5-10. Purchasing Kanban for Suppliers

Buffers

The critical issue for suppliers will be managing the buffer inventory needed to make just-in-time deliveries to their customers who are switching to pull production. You may have noticed in Figure 5-7 that the third type of delivery route includes a supplier distribution center. Unless the supplier also switches to pull production, the demands of their pull production customers will require inventory buffers near customers' sites. *The best way to respond to this will be for suppliers to shift to pull production themselves. If this is done, then the supplier buffer will resemble a supermarket inside a factory,* where kits or bundles of supplies matching kanban container orders are stored and managed according to the rules of pull production.

Key Point

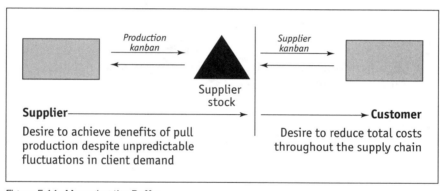

Figure 5-11. Managing the Buffer

TAKE FIVE

Take five minutes to think about these questions and to write down your answers:

1. What are three ways you can begin to synchronize your supply chain to your pull production system?

2. Are there any suppliers you might help to implement a pull production system in their plants too?

Challenges, New Thinking, and Success Factors

As you launch your pull production implementation, remember that the following conditions are important for success:

- Leadership at all levels of the organization must be committed to achieving pull production.

- Resources for adequate training in the new methods and in multi-tasking must exist.

- Behavior must change in one fundamental way: *produce when signaled*, not when material is available.

- Improvements don't have to be expensive, but time to solve problems must be allowed.

- Read the cells *every day* to see what is needed and quickly remove obstacles that are found.

- Align performance measures with the implementation and improvement process.

In Conclusion

SUMMARY

Pull production will increasingly become the reality as you install the kanban system and one-piece flow. One-piece flow is an ideal state that is not always possible, and, perhaps, not always desirable. The important thing is to promote a continuous flow of products with the least amount of delay or waiting in the process. We have said that 5S and quick changeover are prerequisites to pull production. At no time does this become more obvious than in shifting to one-piece flow. If you thought you had uncovered all hindrances, all waste, all retention points and causes of delay in your processes, *shift to one-piece flow and see what happens. You will be amazed at what instantly rises to the surface that you never realized might be causing problems for your operators. This is where the rubber meets the road.*

One-piece flow causes the total uncovering of concealed waste. Some recommend shifting to one-piece flow first, before co-locating equipment and designing cells, and even before implementing quick changeover. This is certainly a good way to uncover process waste. But depending on the type of product and the number of product lines you have, you will probably also have trouble keeping up with orders because of how much waste you encounter in your processes. Therefore, we suggest going in the order prescribed in this book, with one-piece flow as the *last step* in your pull implementation. Even doing it this way, plenty of hidden waste will be revealed. You will run into many obstacles as you work to establish the conditions for flow in your factory, especially resistance in people's minds (including your own) about how this will work. *We recommend starting with the most enthusiastic workshop in your factory and creating a model line.* The model will show everyone in the company how flow production works in a line and what kinds of things it involves. It will give you a chance to work out the kinks in your own understanding of how to implement it as well.

Once you have a successful model line of flow production, you can extend that clearly visible example to the rest of the lines in the factory. This is called lateral development. When flow production is well established throughout the plant you can begin to extend it to your vendors and sub-contracted suppliers. This is called vertical development. *With suppliers, the aim is to reduce the amount delivered, and instead increase the number of deliveries.*

To keep inventories down and lead times short, it will be necessary to significantly increase the number of smaller deliveries. Therefore delivery methods must be changed. If existing delivery methods are used, this will be too costly. Innovation in delivery will continue to be needed from the transport industry in three areas: loading methods for increased product diversification, frequency of deliveries for lower inventory levels and shorter lead-times, and route planning for cost reduction.

Once transport methods meet the needs of pull production, changes in where materials are delivered should be considered. Which part of the factory should take in the delivered items? Where and how deliveries are received can improve the ease and speed of handling materials within the plant and simplify the process of getting them to point of use on the line. *The closer deliveries can be made to point of use the better.* Consider these examples to help eliminate waste in the flow of goods from the delivery sites: *signboards, color coding, product-specific delivery sites, FIFO—first in, first out, and visible container organization.*

Electronic systems are increasingly replacing hand-delivered paper documents to manage procurement, but whether paper, fax, or e-mail, the supplier or purchasing kanban is used to make and track orders. It can also replace delivery vouchers. The kanban system can be used with external suppliers who deliver parts and materials to the factory. Using the kanban system insures that everyone, including the transporter, knows where each item of the order goes. Delivery sites can be indicated on the supplier kanban, as well as the color code indicating the destination line or point of use.

The critical issue for suppliers will be managing the buffer inventory needed to make just-in-time deliveries to their customers who are switching to pull production. Unless the supplier also switches to pull production, the demands of their pull production customers will require inventory buffers near customers' sites. *The best way to respond to this will be for suppliers to shift to pull production themselves. If this is done, then the supplier buffer will resemble a supermarket inside a factory,* where kits or bundles of supplies matching kanban container orders are stored and managed according to the rules of pull production.

As you launch your pull production implementation, remember that the following conditions are important for success:

- Leadership at all levels of the organization must be committed to achieving pull production.

- Resources for adequate training in the new methods and in multitasking must exist.

- Behavior must change in one fundamental way: *produce when signaled*, not when material is available.

- Improvements don't have to be expensive, but time to solve problems must be allowed.

- Read the cells *every day* to see what is needed and quickly remove obstacles that are found.

- Align performance measures with the implementation and improvement process.

REFLECTIONS

Now that you have completed this chapter, take five minutes to think about these questions and to write down your answers:

- What did you learn from reading this chapter that stands out as particularly useful or interesting?

- Do you have any questions about the topics presented in this chapter? If so, what are they?

- What additional information do you need to fully understand the ideas presented in this chapter?

Chapter 6

Reflections and Conclusions

A Summary of Pull Implementation

Five Phases of Pull Implementation

Phase One: Choosing Pull

1. Consider the advantages of the pull system in your production facilities.

2. Consider the status of improvement activities in your production facilities:

   a. What is the awareness level in the principles of waste elimination and continuous improvement?

   b. Do you have on-going team-based improvement activities? Performance measures based on process improvement goals? Visual displays?

   c. Have you implemented 5s? Quick changeover? Cell design?

   d. Do you have a multi-task trained workforce?

   e. Is your union involved and supportive?

Phase Two: Understanding and Preparing for Pull

1. Identify the value stream for your production process.

2. What is the lead-time for each production process?

3. How much time is spent in transport between production processes? What are the distances?

4. How much time is spent in transport between production operations? What are the distances?

5. Where are the retention points in each process?

6. What percentage of the lead-time do the retention points comprise?

7. What percentage of your assets is held in inventory? For how long?

8. How much space is used for storing materials? Parts? WIP? Sub-assemblies? Final products?

Phase Three: Implementing Pull

1. Establish a formal improvement infrastructure.

2. Identify the current production process.

3. Co-locate equipment.

4. Design manufacturing cells.

5. Initiate the kanban system.

6. Shift to one-piece flow.

7. Establish the process for achieving level production.

8. Establish the process for line balancing.

Phase Four: Managing Pull

1. Follow the rules of pull production.

2. Train everyone in the rules and continually improve the kanban system.

3. Set up two-stage pull systems with kit assembly operations and supermarkets for processes that require parts from a large variety of sources.

Phase Five: Extending Pull

1. Establish one-piece flow production.

2. Extend the pull system to your suppliers.

3. Participate fully throughout the implementation process and thereafter to ensure success.

Reflecting on What You've Learned

Key Point

An important part of learning is reflecting on what you've learned. Without this step, learning can't take place effectively. That's why we've asked you to reflect at the end of each chapter. And now that you've reached the end of the book, we'd like to ask you to reflect on what you've learned from the book as a whole.

Take ten minutes to think about the following questions and to write down your answers:

- What did you learn from reading this book that stands out as particularly useful or interesting?

- What ideas, concepts, and techniques have you learned that will be *most* useful to you during implementation of pull production? How will they be useful?

- What ideas, concepts, and techniques have you learned that will be *least* useful during implementation of pull production? Why won't they be useful?

- Do you have any questions about pull production? If so, what are they?

Opportunities for Further Learning

How-to Steps

Here are some ways to learn more about pull production:

- Find other books, videos, or trainings on this subject. Several are listed on the next pages.

- If your company is already implementing pull production, visit other departments or areas to see how they are applying the ideas and approach you have learned about here.

- Find out how other companies have implemented pull production. You can do this by reading magazines and books about manufacturing cells, lean manufacturing, or just-in-time manufacturing, and by attending conferences and seminars presented by others.

Conclusions

Pull production is more than a series of techniques. It is a fundamental approach to improving the manufacturing process. We hope this book has given you a taste of how and why this approach can be helpful and effective for you in your work.

Additional Resources Related to Pull Production

Books and Videos

Waste Reduction and Lean Manufacturing Methods

Hiroyuki Hirano, *JIT Implementation Manual: The Complete Guide to Just-in-Time Manufacturing* (Productivity Press, 1990). This two-volume manual is a comprehensive, illustrated guide to every aspect of the lean manufacturing transformation.

Hiroyuki Hirano, *JIT Factory Revolution: A Pictorial Guide to Factory Design of the Future* (Productivity Press, 1988). This book of photographs and diagrams gives an excellent overview of the changes involved in implementing a lean, cellular manufacturing system.

Shigeo Shingo, *A Study of the Toyota Production System: From an Industrial Engineering Viewpoint* (Productivity Press, 1989). This classic book was written by the renowned industrial engineer who helped develop key elements of the Toyota system's success.

Jeffrey Liker, *Becoming Lean: Inside Stories of U.S. Manufacturers* (Productivity Press, 1997). This book shares powerful first-hand accounts of the complete process of implementing cellular manufacturing, just-in-time, and other aspects of lean production.

Japan Management Association (ed.), *Kanban and Just-in-Time at Toyota: Management Begins at the Workplace* (Productivity Press, 1986). This classic overview book describes the underlying concepts and main techniques of the original lean manufacturing system.

Taiichi Ohno, *Toyota Production System: Beyond Large-Scale Production* (Productivity Press, 1988). This is the story of the first lean manufacturing system, told by the Toyota vice president who was responsible for implementing it.

Ken'ichi Sekine, *One-Piece Flow: Cell Design for Transforming the Production Process* (Productivity Press, 1992). This comprehensive book describes how to redesign the factory layout for most effective deployment of equipment and people; it includes many examples and illustrations.

Iwao Kobayashi, *20 Keys to Workplace Improvement* (Productivity Press, 1995). This book addresses 20 key areas in which a company must improve to maintain a world class manufacturing operation. A five-step improvement for each key is described and illustrated.

The 5S System and Visual Management

Tel-A-Train and the Productivity Press Development Team, *The 5S System: Workplace Organization and Standardization* (Tel-A-Train, 1997). Filmed at leading U.S. companies, this seven-tape training package (co-produced with Productivity Press) teaches shopfloor teams how to implement the 5S System.

Productivity Press Development Team, *5S for Operators: Five Pillars of the Visual Workplace* (Productivity Press, 1996). This Shopfloor Series book outlines five key principles for creating a clean, visually organized workplace that is easy and safe to work in. Contains numerous tools, illustrated examples, and how-to steps, as well as discussion questions and other learning features.

Michel Greif, *The Visual Factory: Building Participation Through Shared Information* (Productivity Press, 1991). This book shows how visual management techniques can provide just-in-time information to support teamwork and employee participation on the factory floor.

Quick Changeover

Productivity Press Development Team, *Quick Changeover for Operators: The SMED System* (Productivity Press, 1996). This Shopfloor Series book describes the stages of changeover improvement with examples and illustrations.

Shigeo Shingo, *A Revolution in Manufacturing: The SMED System* (Productivity Press, 1985). This classic book tells the story of Shingo's SMED System, describes how to implement it, and provides many changeover improvement examples.

Poka-Yoke (Mistake-Proofing) and Zero Quality Control

Productivity Press Development Team, *Mistake-Proofing for Operators: The ZQC System* (Productivity Press, 1997). This Shopfloor Series book describes the basic theory behind mistake-proofing and introduces poka-yoke systems for preventing errors that lead to defects.

Shigeo Shingo, *Zero Quality Control: Source Inspection and the Poka-Yoke System* (Productivity Press, 1986). This classic book tells how Shingo developed his ZQC approach. It includes a detailed introduction to poka-yoke devices and many examples of their application in different situations.

NKS/Factory Magazine (ed.), *Poka-Yoke: Improving Product Quality by Preventing Defects* (Productivity Press, 1988). This illustrated book shares 240 poka-yoke examples implemented at different companies to catch errors and prevent defects.

Total Productive Maintenance

Japan Institute of Plant Maintenance, ed., *TPM for Every Operator* (Productivity Press, 1996). This Shopfloor Series book introduces basic concepts of TPM, with emphasis on the six big equipment-related losses, autonomous maintenance activities, and safety.

Japan Institute of Plant Maintenance (ed.), *Autonomous Maintenance for Operators* (Productivity Press, 1997). This Shopfloor Series book on key autonomous maintenance activities includes chapters on cleaning/inspection, lubrication, localized containment of contamination, and one-point lessons related to maintenance.

Newsletters

Lean Manufacturing Advisor—News and case studies on how companies are implementing lean manufacturing philosophy and specific techniques such as pull production, kanban, cell design, and so on. For subscription information, call 1-800-394-6868.

Training and Consulting

Productivity Consulting Group offers a full range of consulting and training services on lean manufacturing approaches and pull production. For additional information, call 1-800-394-6868.

Website

Visit our web pages at www.productivityinc.com to learn more about Productivity's products and services related to pull production.

About the Productivity Press Development Team

Since 1979, Productivity, Inc. has been publishing and teaching the world's best methods for achieving manufacturing excellence. At the core of this effort is a team of dedicated product developers, including writers, instructional designers, editors, and producers, as well as content experts with years of experience in the field. Hands-on experience and networking keep the team in touch with changes in manufacturing as well as in knowledge sharing and delivery. The team also learns from customers and applies this knowledge to create effective vehicles that serve the learning needs of every level in the organization.

About the Shopfloor Series

Put powerful and proven improvement tools in the hands of your entire workforce!

Progressive shopfloor improvement techniques are imperative for manufacturers who want to stay competitive and to achieve world class excellence. And it's the comprehensive education of all shopfloor workers that ensures full participation and success when implementing new programs. The Shopfloor Series books make practical information accessible to everyone by presenting major concepts and tools in simple, clear language.

Books currently in the Shopfloor Series include:

5S FOR OPERATORS
5 Pillars of the Visual Workplace
The Productivity Press Development Team
ISBN 1-56327-123-0 / 133 pages
Order 5SOP-BK / $25.00

QUICK CHANGEOVER FOR OPERATORS
The SMED System
The Productivity Press Development Team
ISBN 1-56327-125-7 / 93 pages
Order QCOOP-BK / $25.00

MISTAKE-PROOFING FOR OPERATORS
The Productivity Press Development Team
ISBN 1-56327-127-3 / 93 pages
Order ZQCOP-BK / $25.00

JUST-IN-TIME FOR OPERATORS
The Productivity Press Development Team
ISBN 1-56327-134-6 / 96 pages
Order JITOP-BK / $25.00

TPM FOR EVERY OPERATOR
The Japan Institute of Plant Maintenance
ISBN 1-56327-080-3 / 136 pages
Order TPMEO-BK / $25.00

TPM FOR SUPERVISORS
The Productivity Press Development Team
ISBN 1-56327-161-3 / 96 pages
Order TPMSUP-BK / $25.00

TPM TEAM GUIDE
Kunio Shirose
ISBN 1-56327-079-X / 175 pages
Order TGUIDE-BK / $25.00

AUTONOMOUS MAINTENANCE
The Japan Institute of Plant Maintenance
ISBN 1-56327-082-x / 138 pages
Order AUTOMOP-BK / $25.00

FOCUSED EQUIPMENT IMPROVEMENT FOR TPM TEAMS
The Japan Institute of Plant Maintenance
ISBN 1-56327-081-1 / 144 pages
Order FEIOP-BK / $25.00

OEE FOR OPERATORS
The Productivity Press Development Team
ISBN 1-56327-221-0 / 96 pages
Order OEEOP-BK / $25.00

CELLULAR MANUFACTURING
One-Piece Flow for Workteams
The Productivity Press Development Team
ISBN 1-56327-213-X / 96 pages
Order CELL-BK / $25.00

KANBAN FOR THE SHOPFLOOR
The Productivity Press Development Team
ISBN 1-56327-269-5 / 120 pages
Order KANOP-BK / $25.00

KAIZEN FOR THE SHOPFLOOR
The Productivity Press Development Team
ISBN 1-56327-272-5 / 112 pages
Order KAIZOP-BK / $25.00

PULL PRODUCTION FOR THE SHOPFLOOR
The Productivity Press Development Team
ISBN 1-56327-274-1 / 122 pages
Order PULLOP-BK / $25.00

Productivity Press, 444 Park Avenue South, Suite 604, New York, NY 10016
Customer Service Department: Telephone **1-800-394-6868** Fax **1-800-394-6286**

THE SHOPFLOOR SERIES LEARNING ASSESSMENT PACKAGE

Software to Confirm the Learning of Your Knowledge Workers

Created by the Productivity Development Team

How do you know your employee education program is getting results? Employers need to be able to quantify the benefit of their investment in workplace education. The *Shopfloor Series books* and *Learning Packages* from Productivity Press offer a simple, cost-effective approach for building basic knowledge about key manufacturing improvement topics. Now you can confirm the learning with the *Shopfloor Series Learning Assessment.*

The *Shopfloor Series Learning Assessment* is a new software package developed specifically to complement five key books in the *Shopfloor Series*. Each module of the Learning Assessment provides knowledge tests based on the contents of one of the *Shopfloor Series books,* which are written for production workers. After an employee answers the questions for a chapter in the book, the software records his or her score. Certificates are included for recognizing the employee's completion of the assessment for individual modules and for all five core modules.

The *Shopfloor Series Learning Assessment* will help your company ensure that employees are learning and are recognized and rewarded for gaining knowledge. It supports professional development for your employees as well as effective implementation of shopfloor improvement programs.

ISBN 1-56327-203-2
Order ASSESS-BK / $1495.00

Here's How the Learning Assessment Package Works:

1. The employee reads one of the Shopfloor Series books, chapter by chapter. Easy to read and understand, the books educate your employees with information they need, and prepare them for the learning assessment test questions.

2. After an administrator has set up the Learning Assessment software on a computer, the employee can then use the computer to answer a set of test questions about the information in the Shopfloor Series book they have read. The software automatically scores the answers and logs the score into a database for easy access by the administrator.

3. If the employee does not pass the assessment for a particular chapter, he or she can review the material in the book and take the assessment again. (For security, the software selects randomly from three different questions on each topic.)

4. Upon passing the assessment modules for all chapters of the Shopfloor Series books, the employee receives a completion certificate (included in the package) and any other reward or recognition determined by your company.

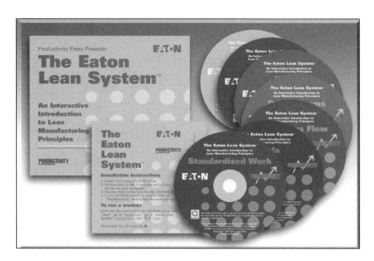

THE EATON LEAN SYSTEM

An Interactive Introduction to Lean Manufacturing Principles

If you're interested in a multi-media learning package, the best one available is *The Eaton Lean System*. Integrating the latest in interactivity with informative and powerful video presentations, this innovative software involves the user at every level. Nowhere else will you find the fundamental concepts of lean so accessible and interesting. Seven topic-focused CDs let you tackle lean subjects in the order you choose. Graphs, clocks and diagrams showing time wasted or dollars lost powerfully demonstrate the purpose of lean. Video clips show real people working either the lean way or the wasteful way. Easy to install and use, *The Eaton Lean System* offers the user exceptional flexibility. Either interact with the program on your own, or involve a whole group by using an LCD display.

This Software Package Includes:

7 CDs covering these important lean concepts!

- Muda
- Standardized Work
- Continuous Flow
- 5S (including 5S for administrative areas)
- Pull Systems
- Kaizen
- Heijunka

Includes in-plant video footage, interactive exercises and extensive simulations!

System Requirements — PC Compatible

Microsoft Windows® '98
8MB Free RAM

QuickTime for Windows 32 bit
16 bit display

The Eaton Lean System
The Productivity Development Team
ISBN 1-56327-261-X
Order EATON-BK / $695.00